THE U.S. NAVAL INSTITUTE ON
THE U.S. NAVAL ACADEMY
THE CHALLENGES

U.S. NAVAL INSTITUTE
Chronicles

For nearly a century and a half since a group of concerned naval officers gathered to provide a forum for the exchange of constructive ideas, the U.S. Naval Institute has been a unique source of information relevant to the nation's sea services. Through the open forum provided by *Proceedings* and *Naval History* magazines, Naval Institute Press (the book-publishing arm of the institute), a robust Oral History program, and more recent immersion in various cyber activities (including the *Naval Institute Blog* and *Naval Institute News*), USNI has built a vast assemblage of intellectual content that has long supported the Navy, Marine Corps, and Coast Guard as well as the nation as a whole.

Recognizing the potential value of this exceptional collection, USNI has embarked on a number of new platforms to reintroduce readers to significant portions of this virtual treasure trove. The U.S. Naval Institute Chronicles series focuses on the relevance of history by resurrecting appropriate selections that are built around various themes, such as battles, personalities, and service components. Available in both paper and eBook versions, these carefully selected volumes help readers navigate through this intellectual labyrinth by providing some of the best contributions that have provided unique perspectives and helped shape naval thinking over the many decades since the institute's founding in 1873.

The U.S. Naval Institute on

THE U.S. NAVAL ACADEMY
THE CHALLENGES

THOMAS J. CUTLER
Series Editor

Naval Institute Press
Annapolis, Maryland

Naval Institute Press
291 Wood Road
Annapolis, MD 21402

Library of Congress Cataloging-in-Publication Data
Names: Cutler, Thomas J., 1947–, editor of compilation. | United States
    Naval Institute, issuing body.
Title: The U.S. Naval Institute on the U.S. Naval Academy : the challenges.
Description: Annapolis, Maryland : Naval Institute Press, [2016] | Series: U.S.
    Naval Institute chronicles | Includes index.
Identifiers: LCCN 2016008331 | ISBN 9781682470237 (alk. paper) |
    ISBN 9781682470244 (mobi)
Subjects: LCSH: United States Naval Academy. | Naval education—United
    States.
Classification: LCC V415.L1 U628 2016 | DDC 359.0071/173—dc23
    LC record available at http://lccn.loc.gov/2016008331

♾ Print editions meet the requirements of ANSI/NISO z39.48–1992
(Permanence of Paper).
Printed in the United States of America.

24 23 22 21 20 19 18 17 16   9 8 7 6 5 4 3 2 1
First printing

CONTENTS

EDITOR'S NOTE

BECAUSE THIS BOOK is an anthology, containing documents from different time periods, the selections included here are subject to varying styles and conventions. Other variables are introduced by the evolving nature of the Naval Institute's publication practices. For those reasons, certain editorial decisions were required in order to avoid introducing confusion or inconsistencies and to expedite the process of assembling these sometimes disparate pieces.

Gender

Most jarring of the differences that readers will encounter are likely those associated with gender. A number of the included selections were written when the armed forces were primarily a male domain and so adhere to purely masculine references. I have chosen to leave the original language intact in these documents for the sake of authenticity and to avoid the complications that can arise when trying to make anachronistic adjustments. So readers are asked to "translate" (converting the ubiquitous "he" to "he or she" and "his" to "her or his" as required) and, while doing so, to celebrate the progress that we have made in these matters in more recent times.

Author "Biographies"

Another problem arises when considering biographical information of the various authors whose works make up this special collection. Some of the selections included in this anthology were originally accompanied by biographical information about their authors. Others were not. Those "biographies" that do exist have been included. They pertain to the time the article was written and may vary in terms of length and depth, some amounting to a single sentence pertaining to the author's current duty station, others consisting of several paragraphs that cover the author's career.

Ranks

I have retained the ranks of the authors *at the time of their publication*. As noted above, some of the authors wrote early in their careers, and the sagacity of their earlier contributions says much about the individuals, about the significance of the Naval Institute's forum, and about the importance of writing to the naval services—something that is sometimes underappreciated.

Other Anomalies

Readers may detect some inconsistencies in editorial style, reflecting staff changes at the Naval Institute, evolving practices in publishing itself, and various other factors not always identifiable. Some of the selections will include citational support, others will not. Authors sometimes coined their own words and occasionally violated traditional style conventions. *Bottom line:* with the exception of the removal of some extraneous materials (such as section numbers from book excerpts) and the conversion to a consistent font and overall design, these articles and excerpts appear as they originally did when first published.

ACKNOWLEDGMENTS

THIS PROJECT would not be possible without the dedication and remarkable industry of Denis Clift, the Naval Institute's vice president for planning and operations and president emeritus of the National Intelligence University. This former naval officer, who served in the administrations of eleven successive U.S. presidents and was once editor in chief of *Proceedings* magazine, bridged the gap between paper and electronics by single-handedly reviewing the massive body of the Naval Institute's intellectual content to find many of the treasures included in this anthology.

A great deal is also owed to Mary Ripley, Janis Jorgensen, Rebecca Smith, Judy Heise, Debbie Smith, Elaine Davy, and Heather Lancaster who devoted many hours and much talent to the digitization project that is at the heart of these anthologies.

Introduction

IT IS NOT SURPRISING that in the many decades since the founding of the U.S. Naval Institute in 1873, there has been a special relationship between that organization and the U.S. Naval Academy, since both USNI and USNA cohabitate along the banks of the Severn River. The Naval Institute was born at the Naval Academy when a group of naval officers gathered to discuss the Navy's problems and to offer solutions, and it has remained there ever since, sanctioned by Congress and embraced by many generations of caring Sailors who see its unique value.

But it is more than a matter of real estate that has led to the natural symbiosis that these two institutions enjoy. As one of the cradles from which officers of the Navy and Marine Corps are nurtured and prepared for the challenges of national leadership, the Naval Academy is rightfully scrutinized, praised, and critiqued by the Naval Institute, whose primary purpose is to make the nation's sea services stronger through the open forum it provides. Over the decades *many* articles have appeared in *Proceedings* and *Naval History* magazines dealing with the U.S. Naval Academy, as well as a number of books published by the Naval Institute Press.

A companion volume, *The U.S. Naval Institute on the U.S. Naval Academy—The History*, records the Academy's founding and its subsequent development, but this edition of Chronicles presents a number of selections from that large catalog of Naval Institute offerings that deal with the challenges that this unique institution has faced over more than a century and half since its founding. The Naval Academy is not your typical hall of higher learning, and its iconoclasm, coupled with the fact that it is largely funded by the nation's taxpayers, makes it vulnerable to frequent questioning, if not outright attack. It is not always understood by those who are footing the bill, and its long tenure ensures that it has had to weather the storms created by changing conditions in the nation and the world.

And that is where the Naval Institute comes in. For most of the Naval Academy's existence, the Naval Institute has provided an open forum where questions can be asked—freely and without undue influence (despite their common residence on the banks of the Severn)—where answers can be offered, and where useful information can be disseminated to explain the Academy's functions and its importance.

Sometimes these offerings merely enlighten outsiders and remind insiders of the unique character and history of this school where the goal is to merge the best of Athens and Sparta. At other times these writings offer helm orders designed to keep this vital ship on the proper course. And occasionally there are existential challenges that any worthwhile endeavor must be prepared to endure. As readers will no doubt see in these pages (which represent only a portion of the overall corpus preserved in the Naval Institute's archive), the ongoing process of edification and of challenging dialog is a healthy—if sometimes painful—process of symbiosis and disputation that ultimately serves both institutions well and the nation even better.

Prologue:
"A Perfect Form One"

Commander Jim Stavridis, USN

U.S. Naval Institute *Proceedings*
(October 1995): 45–47

A "FORM ONE," according to the book of naval signals, is a formation of ships at sea maneuvering in a column, where one vessel follows behind another, taking its appointed place and sailing in the wake of the warship just ahead. The long gray hulls are a beautiful sight, gliding one after another through the churning sea, creating a stirring example of the timeless power and grace of the naval service under way. But the Form One also is a striking symbol of another sort. It perfectly evokes the faces of the Naval Academy, that long line of men and women, one following another, who serve their country so well both afloat and ashore.

On their initial day at the Naval Academy, young men and women take their places in the first of thousands of formations, their faces bright and excited, representing every race and religion and background. They are a gift of inestimable value given by the United States to the naval service each year. The new midshipmen who arrive at the Naval Academy each summer are untouched faces, upon which so much eventually will be written, both at Annapolis and beyond. From their numbers will come admirals and generals, infantry officers and naval aviators, bridge watchstanders and platoon commanders—as well as many distinguished civilians who leave the service after their obligations are fulfilled.

Their faces will witness countless sunsets on the deep, rolling sea; they will take the watch on a destroyer's bridge a thousand times and more, searching the distant horizon for barely seen contacts; they will lead companies of Marines down dusty streets into danger and adventure; they will dive nuclear submarines under the polar ice and fly the fastest jet aircraft to the highest places in the sky. For much of their lives, those faces will be turned away from their homes as they stand watch in the long, hard service of their country, defending the hope and the promise that defines the United States.

In every issue of *Shipmate*, the Naval Academy's alumni magazine, each class has a monthly column full of news and photographs of the graduates. Start at the back, with the youngest and newest graduates. Their faces stand out in their youth and energy, unlined and beautiful, bursting with promise. Living their lives so close to the flame, they are sure they are indestructible, with countless options ahead and so little in the way of history dragging behind. They are, in the words of former Secretary of the Navy Sean O'Keefe, as full of light as the sun, as full of grace as angels.

Turn the pages.

The years flip by, and the faces age. Wives and husbands and children appear. First deployments are completed, and airmanship and seamanship are mastered; these are the building years of service as a junior officer. The first faint lines can just be seen on those faces, beginning around the eyes that have seen so many sunsets at sea, so many hot summer days in the deserts of Arabia, so many long patrols over the choppy Adriatic. The faces deepen and begin to develop new expressions: gravity, seriousness, maturity. The burden of the years begins to show.

Turn the pages yet again.

Responsibility, accountability, command—important things that again change the faces in subtle ways. Gray appears at the temples, hairlines recede, and lines deepen. The cares and concerns of a demanding life, much of it spent at sea, begin to make themselves felt in those faces.

Sons and daughters grow, and soon the first child of a classmate is entering as a plebe at Annapolis. There are commanders and colonels, flag and general officers—some faces are moving along at higher and higher speed, headed toward yet more demanding tasks.

Turn more pages and suddenly the burdens begin to change and lift: retirements, second careers, transitions. Soon the first grandchild enters the Naval Academy, a distant, blurry repetition of the one that walked into the Academy so many years ago. Travel and reunions, Florida condos and tennis matches, sailboats and golf villas—the rewards of a life well led appear on those pages. The faces relax and smile, even as the accumulation of life's lines continues to build on faces increasingly full of character and experience.

The last pages.

Fewer faces now. Reports of illness, condolences and obituaries, memorials and bequests, the graceful conclusions of orderly lives are documented; and then abruptly the faces stop—the final watch stood, the last log signed.

There is a stately rhythm to it all. Each stage has a different feel to it, a different defining quality that is caught in the faces of the men and women living it.

What does it all mean, this parade of changing faces?

Caught in the pages of *Shipmate*, the faces of the Naval Academy have a meaning far larger than the institution itself. The faces show what is really important in life: families, friendships and humor, service to a higher ideal, the idea of continuity, and—above all—the naval service itself.

First, those faces clearly demonstrate that families are the heart of the matter. Each young face comes to the Academy from a family that supported and cared for them. And as these young men and women leave Annapolis and pass through the building years, they gather their own new families about them. The career they have chosen is hard on those families, yet many children of graduates pay their parents the ultimate compliment when the time comes to choose their life's work and say, yes, I too will serve—often with the determination to attend the Academy themselves.

That repetition of the cycle, the long steady sweep of an endless Form One, is fundamental to the Naval Academy. And, more important, it is at the heart of the Naval Academy's value to this country and the naval service. The next generation must choose freely to come and serve and to be part of that long formation of faces—and only families, with their love and support, can encourage and achieve that particular progression of face after face.

Second, those faces demonstrate that friendship and humor matter deeply. Bonds are built at the Academy, and the strength of those ties are shown on page after page, on face after face. The laughing ensigns and second lieutenants posing in the sunny hours after graduation become the smiling captains and colonels 25 years later, and the same friendships and humor will hold the naval service together from the corridors of the Pentagon and to the piers of Port-au-Prince and the runways of Mogadishu.

Third, the faces show that the truest calling is service to a higher good. Like the faces in Memorial Hall at the Academy, these are the faces of good men and women who often have paid a high price for the right to serve. Some faces disappear early indeed, lost in combat, fallen in the accidents of flight, or swept over the side of a ship riding harshly in a cruel sea. Those are the faces of young men and women who will never find their way to the final pages of *Shipmate*, lost hearts swept away from the rest of us by the dangers of the lives they chose. But those faces never are truly lost if they are preserved and remembered by their comrades and friends. They are forever a part of all that the faces represent, a part that is remembered in the memorials and monuments that are such a quietly important and powerful part of the Naval Academy grounds.

Fourth, those faces tell us that there is a comforting continuity to life. In those faces that emerge from the Naval Academy, year after year, it becomes clear that each generation will indeed take its stand and face great issues; then surrender its place and move on. As said in Ecclesiastes, "One generation passeth away, and another generation cometh, but the earth abideth forever." So it is with the naval service and the long Form One.

We all learn that no one among us is indispensable, that there will be both victories and defeats, promotions won and lost, choice assignments gained and missed—all vitally important at the moment. But in the end, the progression of faces shows that the world moves on, effortlessly and seamlessly, catching up in its movements the best each of us has to offer, hopefully taking what is good and true and passing some sweet part of it all along to the next generation. There is high comfort in all of that, in the long steady sweep of face after face, generation following generation, like ships in the wake of a guide, steaming smoothly toward a distant horizon.

Finally, that long parade of faces helps to show that what is important is not the precise nature of service, or lists of accomplishments, or final ranks attained. What matters is the service itself, both the act of service and the privilege of performing it in the naval service. Knowing this helps us to take ourselves less seriously.

Sooner or later, everyone is denied something that they think is theirs by right—a perfect assignment, an early promotion, screening for command, selection for war college. But everyone, even the most successful admirals and generals, eventually will arrive at the point where the Navy or Marine Corps says, enough, you have served well and long but now you are done. Understanding the meaning of the long progression of faces can make that moment understandable, if not entirely pleasant. It leads to acceptance and to satisfaction with a job well done and a career well led. In a word, that parade of faces provides perspective.

Graduates of the Naval Academy will have their turn to move through the pages of *Shipmate*. Some will leave the service early to pursue other challenges; some will die young on distant missions; some will command great fleets and forces; and some will have a single command or none at all. Yet for each, the key is balance and perspective, an enjoyment of the moment, a love of friends and families, and an abiding vision of service to a higher ideal.

The faces of the Naval Academy form a family. They are all part of the long formation. They are not all happy in the same ways, nor do they serve in the same ways, but there are two things that each can share. First, all serve—in different ways, for different durations, with different results—but all serve. Second, and more important, the choice to be happy—to find satisfaction with service—lies within each of them. It is the key to everything. Satisfaction with service—a simple and timeless idea—is the most important lesson of the perfect Form One. It is the most fundamental meaning of that long progression of lives and adventures that is composed of the faces of the U.S. Naval Academy.

Commander Stavridis is the commanding officer of the USS *Barry* (DDG-52).

1 "What's the Matter with the Naval Academy? A Plea for a Five Year Course"

Lieutenant Commander A. W. Hinds, USN

U.S. Naval Institute *Proceedings*
(1912): 187–94

THERE IS PROBABLY no school in the country with a better reputation for thoroughness than the U.S. Naval Academy; and there are many reasons why the Academy should have this fair reputation. The instructors are conscientious—both the civilians and the naval officers. The sections consisting of from 9 to 12 midshipmen are small enough, ordinarily, to allow for individual instruction in addition to the assignment of marks for gradation. Compared to other schools the sections here are only about one third as large.

The arrangements for the midshipmen are almost ideal. The quarters are comfortable; the food is good; the study and exercise hours are remarkably well balanced; during study hours there is no interference by room to room visiting; there is not the slightest ground for the criticism of the late Mr. Crane on all college courses, i.e., that a large percentage of college undergraduates spend their time getting drunk.

It would be a difficult matter to suggest any method of teaching better than the one that has been used here, with success, for many years. Lessons are assigned and the midshipmen dig them out for themselves. Day after day they recite on these lessons—tell what they have learned about them and receive a mark for the day's work. This gives them confidence and teaches them to rely on their own efforts. At the end of a month an examination brings out what the midshipman have retained of the month's work and the semi-annual and annual examinations show whether they have kept in their minds the general principles covered during the term's work. The system of teaching used at the Naval Academy is used by very few other schools in the country—in fact West Point is probably the only other school using it. The reason of this is that at the other institutions of learning the sections are too large and the instructors too few to allow much individual instruction or to allow for individual recitations. The number of instructors provided for the school is of course a function of the money provided for running expenses; the number here is, ordinarily, sufficient for the needs of the school. Where there are 25 or 30 students in a section the only feasible way to cover the subject matter is to have lectures, i.e., have the instructor do the reciting, and find, from examinations, what the students have learned. There are two main objections to the lecture system: (1) There is a certain amount of inattention even at the most interesting lecture; at some time during the talk the thread will be lost, and, even though this may be momentarily, the value of the lecture will be decreased by reason of such loss. (2) It is a bad policy to work out the difficult problems for a student; he will remember much longer if he is compelled to work out everything for himself. It is very probable that if the other schools of the country had enough instructors to keep the sections small they would use the same system of teaching used at the Naval Academy.

Criticism Creeping In

Notwithstanding the fact that the methods of teaching used here are good, and that the midshipmen probably use more of the four years at the Naval Academy to advantage, on an average, than is used in other schools, criticism of the Academy and its graduates are creeping in— from the fleet and from other sources. A straw vote taken here would show that the majority of the instructors and heads of departments think that in preparing them for their life work we are not giving the midshipmen the best course we could give them.

The main trouble with our course as it now stands is that it is one year too short. Twenty-five years ago the principal studies at the Naval Academy were mathematics, English, physics and chemistry, seamanship, navigation, and modern languages. Marine engineering and ordnance required only a fraction of the time they now require.

At that time the course was well covered, for there was great competition for the comparatively few commissions issued to each class. The time taken to graduate was the same as it is now—four years. During the twenty-five years some of the studies like mathematics, navigation, physics and chemistry, English, and modern languages have changed very little. Other studies like marine and electrical engineering and ordnance and gunnery have changed greatly. The changes in the last named have all been additive. It seems fair to say that all three have doubled the amount the midshipmen have to learn in order to thoroughly understand the subject; and yet, with the entrance requirements very close to what they were then we are trying to cram all this additional learning into the heads of the midshipmen in the same length of time as was taken twenty-five years ago. A comparison of the number of two hour periods spent on each subject in 1889–1900 and 1911–1912 is interesting.

	1899–1900	1911–1912
Mathematics	496	442
Mechanical drawing	208	112
English	208	192
Languages	224	224
Seamanship	65	128
Navigation	176	176
Physics and chemistry	140	96
Electrical Eng.	110	192
Marine engineering and naval construction	304	320
Ordnance and gunnery	96	160

This table shows that mathematics has lost 54 periods, physics and chemistry 44 periods, and mechanical drawing 96 periods during the past ten years.

The loss in mathematics is the one that does the heaviest damage; it hurts the courses in physics and chemistry, in navigation, in ordnance and gunnery, and in marine and electrical engineering—practically all other subjects. On account of lack of time and lack of practice in mathematics, the students, at the present time, take up new subjects poorly equipped. They cannot perform the ordinary operations of algebra and trigonometry rapidly and accurately, and instead of spending their time and energy on new work they have to spend it in reviewing mathematics in order to apply it intelligently to the allied subject.

The loss in physics and chemistry is also a drawback. From a study of the courses at West Point, and at other schools, it appears that they spend much more time on these subjects than is spent at the Naval Academy. A knowledge of both physics and chemistry is needed in a successful naval career. We are constantly called on to use the general principles learned from them. A knowledge of chemistry is absolutely essential in the engineering of the present day. A midshipman must have a reasonable understanding of the science of physics before he can successfully take up the study of marine and electrical engineering. Laboratory measurements, experiments and demonstrations as well as recitations are necessary. The Naval Academy has excellently equipped chemical and physical laboratories, but insufficient time is allowed for their use. No college or scientific school expects a student to acquire a working knowledge of these subjects in less than three times the number of periods allotted to the study of physics and chemistry at the Naval Academy. The study of sound, light and heat is closely allied to that of mathematics; it requires the midshipmen to think and, after all, the practice of reasoning is one of the best features of the Naval Academy education.

If enough time were given to mathematics to get it thoroughly ground in so the midshipmen could apply it successfully to the allied subjects, the time allotted to marine engineering and to navigation would be sufficient as it stands. This does not take account of the loss of time in mechanical drawing which is one of the branches of marine engineering; 112 periods is not enough; 160 periods would be nearer the necessary time. Incidentally, there is good authority that the army and outside concerns are glad to get our graduates because they are good in mechanical drawing. A blueprint of a piece of machinery is not an enigma to them. This will not hold true when we begin to graduate the midshipmen with only 112 periods spent on mechanical drawing.

In addition to more time devoted to mathematics, and to physics and chemistry, more should be given to electrical engineering and to ordnance and gunnery.

It is pretty generally admitted that naval officers in our service are weak in electrical engineering. There are too many things on board ship that depend on electrical gear to allow this condition to exist. Lighting the ship, its ventilation, working the turrets and guns, supplying ammunition, interior communication, fire control, in fact the whole efficiency of the ship as a fighting machine is so vitally affected by the efficiency of the electrical installation that we cannot afford to graduate midshipmen and leave them in an admittedly weak condition as far as electrical knowledge is concerned.

Ordnance and gunnery is in the same plight—there is simply not enough time to cover the subject properly. The head of the department states that it would require one-third more time than is now assigned to that department in order to send the midshipmen away from here well qualified.

Comparison with Other Schools

During the past four or five months the writer has obtained data from many of the leading colleges of the country. While the information was very full, it is rather hard to compare these institutions with the Naval Academy on account of the different methods of teaching and marking.

In making the comparisons engineering courses were selected as they bring in mathematics and engineering, and it was thought that allowance could be made for the additional subjects taken at the Naval Academy, subjects which are strictly professional and which are not taken at other schools. While we did not get everything sought for from the information sheets from these colleges, the following facts were established beyond question: (1) The entrance requirements are higher at all the colleges heard from than at the Naval Academy. (2) The mathematics at the academy is about the same as that of the other colleges. (3) The courses in engineering are about equally difficult. (4) Not one of the colleges has a course shorter than four years. (5) Two of the colleges, although having more rigid entrance requirements than the Naval Academy, do not graduate their students in engineering in less than 5 years. (6) Ten of the eighteen used for comparison have additional courses of from one to three years in engineering—this means after the preliminary four years of work.

We have at the Naval Academy the brothers of these college men. The average brain power here ought to equal that at the colleges. How then can we cover thoroughly the same ground in four years that they cover in an average of more than four years if the college students are better informed when entering college.

But we do not cover the same ground—we cover much more. The mathematics and engineering are about the same. The English and physics and chemistry are perhaps a little easier than in the engineering courses at other schools. But many things are covered here that engineering courses in colleges never touch, as for instance—navigation, astronomy, modern languages, seamanship, international law, ordnance and gunnery. How can we ever cover this extra work in less time than the other schools take when their men are better prepared at entrance. There is at hand a letter from a Naval Academy graduate so apropos that I must quote an extract from it: "I am a graduate of the U.S. Naval Academy, retired, and have been teaching twenty-five years. There is no doubt that your course should be one of five years for the best results to the student. Time is required to do good work anywhere. To teach students fundamental principles so that they cannot forget them even if they try; to teach students to do things, to originate, instead of copying, requires time."

Lack of Competition

Competition at the Naval Academy was killed ten years ago with the advent of the big classes. It seems unnecessary to call attention to the value of competition. The wonderful improvement in gunnery and in engineering in the navy are examples of the result of competition. In the days when the commissions per class were but few, there was a scramble for them and the best man got them. The commissions were the reward of hard work and well cultivated intellect, and the graduates who got them were the thinkers of the class.

With the large classes came the need of a greater number of officers. We had to have them to man the ships. The midshipmen knew this and studied less. The marks went up; 2.5's gave way to 3.0's, for they all had to be graduated.

All this resulted in what was near kin to a union. The midshipmen realized that if the bright ones studied hard the ones who were "wooden" could not get through. It became bad form to "study out of study hours," and why not? It is against human nature to do more than one needs to do. The plane of marks was necessarily shoved up in order to supply our own needs, and, once up, it became difficult to lower. It is not my desire to go on record that these men will not make good officers. I trust they will, I believe they will if they work hard. This much, however, is certain; they must cultivate out in the service the thinking power that they should have more nearly perfected while at the Academy.

Next year, unless the laws are changed, the large classes will cease to exist. This will be a bad thing for the Academy and for the service at large. The classes should be kept as large as they are now, but only the number needed to fill the vacancies should be commissioned. This would give a lot of education at the expense of the government, but the money would be well spent. Furthermore, the students who spend 5 years at the Academy, even though not commissioned in the navy at graduation, would be of great value in war time.

The Bureau of Navigation could decide how many are needed each year. Out of an entering class of say 300 there might be 100 or 120 to commission, and competition, one of the most powerful incentives ever used at the Naval Academy, or anywhere else, would help work out the Naval Academy problem.

Age of Entering

The age of admission and the entrance requirements should be low for many reasons, among which are: 1. Algebra and geometry form the foundation of all mathematics; if those two subjects were taught here we could be sure the midshipmen have the proper foundation to build on. 2. The Naval Academy would get a more ambitious set of boys. Suppose, for instance, the lower age limit to be 18 years; most boys of that age who are ambitious have decided what they are going to do in life, and even at 17 many of them have started the education that is to fit them for their life work. 3. The younger the midshipmen are the more impressionable they are and the easier it is for them to drop into the "Customs of the Service." 4. If taken, say on an average of two years younger, they could put in the additional year required for a proper education and still become commissioned officers two years and a half younger than they do now. The age of entrance is now 16 to 20, a mean of 18 years; add 6 years for the present course (counting the two years at sea) and the average age for commissioning is 24 years. The Smith Bill makes the entering age 15 to 18, a mean of 16.5 years; add to this the 5 years which are needed here and there results an age for commissioning 21.5 years, which is two and a half years less than the present age when commissioned.

Summing Up

A five year course would give sufficient time in mathematics to properly fit the midshipmen for studying the other subjects taught here; it would give sufficient time in physics and chemistry, in ordnance and gunnery and in electrical engineering. It would give a total time of one year on summer cruises, which ought to furnish enough sea experience to justify a commission on graduation.

And if the congressmen and senators are allowed, after 1913, the same number of appointments they now have the percentage of commissions issued to a class at the end of 5 years would be small enough to keep alive the keenest kind of competition.

2 "The Naval Academy of Today and Its Mission"

Lieutenant Commander
Mahlon S. Tisdale, USN

U.S. Naval Institute *Proceedings*
(March 1923): 453–63

The Mission

"To mould the material received into educated gentlemen, thoroughly indoctrinated with honor, uprightness and truth, with practical rather than academic minds, with thorough loyalty to country, with a ground work of educational fundamentals upon which experience afloat may build the finished naval officer, capable of upholding, whenever and wherever may be necessary, the honor of the United States; and withal giving due consideration that healthy minds in healthy bodies are necessities for the fulfillment of the individual missions of the graduates; and that fullest efficiency under this mission can only be attained if, through humane yet firm and just discipline, the graduates carry into the Service respect and admiration for this Academy."

THE NAVAL ACADEMY HAS PASSED through many physical transformations since its inception in 1845, and it requires some daring, in these days of progress, to attempt to visualize the future. But even with the reduction of armaments, it is reasonable to assume that the Naval Academy of today will be the Naval Academy of the future—for several decades at least.

In the evolution from which resulted the present Naval Academy, the greatest single increase in capacity came during the World War when the plant was increased to its present size. By duplicating the wings of Bancroft Hall, the quartering capacity was about doubled. The Hall can house, in its four wings, two thousand forty midshipmen, with two in each room. During the war the appointments were increased to five for each Congressman, with twenty for the President and one hundred for the Secretary of the Navy. As a natural result last year saw a regiment of 2,431, the largest in the history of the Academy. There are 2,396 at present.

Such regiments are too large. They necessitate berthing three midshipmen each in some of the old "Youngster" rooms, and as many as four in some of the "first class suites." Manifestly such crowding, though necessary in but a small percentage of cases, is unsatisfactory. Last year a bill was introduced in Congress to reduce the number of appointments to two for each member of Congress, and proportional reductions among the other appointing powers. This bill has not become law. A happier choice in reducing would be four. There are several good reasons for not reducing below that number. Such a drop would furnish a regiment of about two thousand which would permit the existing plant to be run at maximum efficient capacity. The attrition in commissioned personnel in these times of treaty navies is high. This year it may run as high as two hundred fifty. If the Service is ever to fill existing vacancies, and keep step with attrition as well, we will need about four hundred graduates each year for some years to come. Further, by graduating more than are needed for the Service, the navy can select only the more apt for commissions. Perhaps the Government cannot afford to run the Academy at a maximum, solely to permit choosing the graduates, but there is another side.

The graduates who are returned to civil life will be the nucleus of the finest sort of a reserve. The opportunity for an education will be granted to many young men who might be denied an education otherwise. Four years here cannot fail to instill much Americanism, as well as good character, into the graduates. They will be returned to civil life better citizens in every way. The overhead expense increases only moderately per capita, with a regiment of two thousand, as opposed to one of fifteen hundred. Naturally the pay and the cost of the rations depend directly upon the number of midshipmen. The money could be spent much more unwisely.

These reasons are sound, and contain most of the arguments customarily advanced. Little opposition to them has developed, and there has been no refutation. The Government has many farms for moulding and developing live stock. Surely a school for the moulding of humans into good citizens is a wise investment.

As has obtained for years, the midshipmen are drawn from all walks of life. Yet even the poorest judge of character cannot fail to see that the average candidate compares most unfavorably with the average first-class man.

Recent years have witnessed an increasing interest in what one might call practical or applied psychology. The war focused thought on this feature of leadership. No officer will deny that the ability to lead men is an essential characteristic of a good officer. Yet for years each coming generation has had little guidance in this regard. For our particular form of career, we have had instilled all sorts of knowledge about all sorts of machines, except the human machine. And without the human machine, of what avail are all of the others?

The press of war saw all kinds of men commissioned. Some with previous experience as executives, but more without it, became officers. These men, fired by the enthusiasm of war, patriotism, and personal ambition, sought written guidance—only to find that in the Service experience is, in truth, not only the best teacher, but virtually the only teacher.

Training had to be conducted largely by word of mouth and by example. There were plenty of excellent works on war and similar subjects, but few on the fundamentals of leadership. All of this brought the realization that the graduates of the Academy were being sent into the Service, with possibly the essentials of command indoctrinated into them, but with little tangible advice or instruction in those practical methods, which the experience of generations had shown to be best suited to the situations a junior officer encounters afloat.

It is perhaps true that leaders are born and not made. But there are various degrees of skill among leaders. The Napoleons may perhaps have the God-given gift, but we cannot all be Napoleons. Nor can we wait for the presentation of the gift. Few there are who have it. Yet the rest of us must struggle along as best we can. We are in a career which is primarily one of leadership. The Service standards are of the highest. To meet this situation, a course in leadership has been introduced into the Naval Academy curriculum. And for the same reason, the mission of the academy has been reduced to writing, in the belief that a visible mission will crystallize thought on those features of the academic life which have the greatest bearing in fitting men to command.

Recent developments in warfare such as aviation, gas, radio, and communications have demanded, or are demanding, a place in the curriculum. But it is difficult to find the necessary time. The pace is already fast. The departments have been re-working their respective courses, cutting and pruning where possible, to make room for the newer subjects. The mission is to teach fundamentals, which grants plenty of authority for cutting—but there is much that cannot be cut. The new developments have caused no reduction in the importance of the old subjects. Most of them are in the line of additional information required of the naval officer, rather than a replacing of the old knowledge. Radio, for example, is of a continually increasing importance, but it does not relieve the officer of today from having a knowledge of signals.

To meet this difficulty, there appear to be only two courses open. Either we must come to the five-year course, or the entrance requirements must be so increased that a candidate will have had one year of college work prior to admission. The latter plan seems impracticable for several reasons. It would increase the age of admission, even though the actual requirements might remain unchanged. It would prevent many worthy young men from qualifying for the academy. It would probably not meet with the approval of the Congressmen, who are privileged to appoint the candidates. The present entrance requirements have, in the press from time to time, already been classed as difficult. As a matter of fact, they are not difficult, as the results would indicate. The entrance requirements for classes entering in 1923 and subsequently have been increased, but only up to college entrance requirements.

Now coming to the mission, it contains nothing new, but does express in black and white what the Academy is working for. This differs little from what has been taught for years, but it permits all to see at a glance what a high standard the service expects of its officers; and that, while scholarly attributes are essential, unimpeachable character is vital.

To Mould the Material Received into Educated Gentlemen, Thoroughly Indoctrinated with Honor, Uprightness and Truth

Character permeates the entire mission. I cannot better express the necessity than to quote from a letter written last year by the present Superintendent to the Secretary of the Navy:

> High standards of character are vital in a military organization. Men in other walks of life may trifle with the truth in everyday affairs and suffer little as a consequence. We cannot be so tolerant. The fundamental of an officer's value to the Service is his trustworthiness.

No one in the Service can amount to much without having pride in his calling, and pride in his own particular assignment. This is one reason why the mission sets such a high standard. Men are prone to be what their superiors think they are. The mission directs the thoughts of the young gentlemen of the regiment, as well as those of the officers, toward a target. The target involves honor, uprightness, integrity, and many other essentials. As the midshipmen respond to the challenge and indulge in introspection, each will see some of the dross not visible to others, and a thoughtful consideration of self, compared to the ideal which the Academy demands, gives a determination to meet the test. When it has been met, there comes the pride of success. Custom and tradition have always held ours to be an honorable calling. Ever since there have been navies, officers and men have been proud to wear the blue. A high standard set—and met—makes for satisfaction in one's work.

Pride is allied with self respect, without which no officer or man can succeed. A midshipman's word is not questioned. He is taught that he is expected to tell the truth; that less than the truth shows him unfit; and, so far as the authorities are concerned, that what he says is the truth. This gradually cements into his character not only the habit of truth, but also pride that he is trusted, and his self respect is strengthened accordingly.

When the occasional misfit does enter, sooner or later he shows his true colors. The standard punishment for any offense of evasion of the truth, actual untruth, or other offense involving the honor of the individual or of the Service, is dismissal. The honor element of naval character must be unimpeachable.

It goes without saying that the academic course must be maintained at a high standard. There are at least two reasons for this. First, because academic education trains the mind to think, to reason logically, and to estimate correctly the countless situations which a naval officer is called upon to handle. Second—and this reason is not so widely understood—because a strenuous academic course instills a fundamental of naval character into the midshipman. This is the habit of working, and of working "on one's own." Our section room methods do not meet with the universal approval of the scholars of the country. Some of them prefer the lecture method. But there is no denying that four years spent in digging things out for one's self breeds independence, and teaches the young gentlemen of the regiment to work, and to work hard.

Marks are a necessary evil. One might almost say that they are the curse of the course. They must exist not only to stimulate effort, but to permit some sort of reasonably just arrangement of the graduates on a seniority list. Any other scheme would make the arrangement haphazard.

A primary requisite in the character of the successful naval officer is obedience. One who cannot command himself cannot command others. Hence, for four years the midshipmen are taught self-discipline. Many of the academy regulations seem unnecessary to the thoughtless. But all have a purpose. The American character is essentially one of action. This action in the younger generation is frequently without regard for the rights of others; and is all too often without regard for the consequence. In short, obedience is not one of the prominent characteristics of the average young man. No officer can be successful who is disobedient. This involves loyalty, subordination, and several of the other recognized attributes of one who would succeed. Of course the primary function of most rules is for the orderly carrying out of a large and intricate organization, but regardless of the reason for the regulations, it is a fact that they do teach the novice that there are certain things that he may not do. They form the entering wedge, in the training of men for a life of self-sacrifice in duty's name. The orderly arrangement of the day teaches promptness, loyalty, attention to details, subordination to superiors, and obedience in general. An important part of the training in obedience is military drill—the so-called "drills of precision." They form the habit of obeying an order exactly, at a definite time, and without question. Infantry drill, so far as a necessity for the knowledge of the subject is concerned, is not as important as is gun drill. But as a moulder of character, its value is incomparably greater.

Another important point now receiving much attention is how best to stimulate the efforts of the midshipmen. Every regiment has had a minority who shout vociferously that all they want is a "little old two-five." While this claim is most often made by those incapable of garnering more, such statements to a certain extent are harmful to the rest. Most of us feel a wish that we could do it over again, and the feeling always carries the mental qualification that "I could pull my class standing up, now that I realize what standing means." In other words each of us has admitted, at least to himself, that his present standing does not represent his best effort. The word "greaser" still carries the old stigma, and the midshipmen of today are no different from those of yesterday. To urge the regiment to greater effort, the attention of all hands was called to the value of class standing, and its importance in their career. Forming the habit of doing one's best is much more far-reaching than simple class standing; for learning to do one's best, on all occasions, is a fundamental of naval character.

In a further attempt to stimulate effort along those lines an order was published, explaining the value of service reputation, and encouraging the midshipmen to work now toward that goal; stating that the Superintendent will place a letter of commendation on the records of "those six midshipmen who, at the end of the academic year, will have contributed most by their officer like qualities and positive characters to the development of military spirit and loyalty within the regiment."

It has been said that too many first-class men leave the Academy without having heard the sound of their own voice raised in command. The midshipmen officers received plenty of training, while the "clean sleevers" had to go on board ship into a division assignment, with no experience in command beyond handling a section during a periodic detail. To equalize the opportunities of all first-class men—for after all, our major simple mission is to command—the present system of changing the assignments each month was placed in effect. All first-class men now wear a single narrow stripe around the sleeve; and in addition, the midshipmen officers wear the appropriate number of short stripes just above the class stripe. For the last three months of the year, selections are made from the entire first class, having in mind the record of each midshipman in his previous detail. The best are selected, and they wear the appropriate insignia of rank, the customary five stripes, four stripes, etc., all the way around the sleeve, in lieu of the class stripe. Each midshipman, under this system, graduates with the experience of having commanded a squad at least. This breeds confidence, in proportion to the responsibility which the size of the command has carried. Confidence is another fundamental characteristic of the efficient officer.

As another important step toward the moulding of character, an intensive educational campaign is being conducted to counteract the ease with which lenders of money, or credit, prey upon the midshipmen. The midshipmen have been told repeatedly, both orally and in writing, of the menace carried by the debt habit. An effort has also been made to interest their parents in this feature. One large firm has voluntarily agreed to solicit no life insurance from midshipmen. A Baltimore paper has published an editorial "Rally around the Admiral," commenting favorably upon the efforts being made to stamp out the debt habit. If the midshipmen can be made to see the foolishness of spending considerable sums on "grad" terms, they will not only have avoided forming a bad habit, but will join their first ships more cheerfully than where, as formerly in some cases, a year's pay is obligated to a human vulture.

With Practical Rather than Academic Minds
The midshipmen are taught to seek responsibility, and are given the necessary knowledge, that they may not fail to measure up. The course is essentially practical. An officer handling huge ships, big guns, intricate machinery, and high speeds, must have common sense, a quick acting mind, and the requisite knowledge.

With Thorough Loyalty to Country

Naval Tradition has an important function in developing loyalty to the Service, as well as to the Country. Young officers, though they may have unknowingly absorbed much of it, may scoff at tradition as one of the hobbies of the senile. Thinking men will admit, however, that tradition in any vocation makes for greater efficiency. The young gentlemen are at a most impressionable age, and despite the penchant that youth has for jeering at those who have gone before, it is certain that some inspiration must come from a knowledge of the successes of their predecessors in the Navy. What young man—mayhap he is studying the history of the World War—can look upon Cribble's "The Return of the Mayflower," depicting the entrance into Queenstown of our first destroyers in 1917, without a thrill of pride, and without experiencing a determination to be ready himself when duty calls?

Who of us can look upon the tablet in Memorial Hall, dedicated to those who lost their lives after a voluntary 1,500 mile voyage in an open boat, to bring succor to their comrades of the shipwrecked *Saginaw*, without at least wondering if we can meet the supreme test so courageously? Does not the tablet to Lieutenant Stanton F. Kalk, who surrendered his place on a life raft, after the torpedoing of the *Jacob Jones* in 1917, and in consequence lost his life that an enlisted man, a non-swimmer, might live, conjure up visions of a young officer meeting his Maker manfully, in accordance with "the best traditions of the Service"? And does not the gazer have inculcated just another bit of naval character, to help him in his own hour of trial, when duty will call, and when he must not be found wanting? And so on through the many books, portraits, monuments, and tablets, which the Naval Academy preserves to assist in the character moulding of its youth. A brief history of the more important memorials has been printed, and a copy has been furnished each midshipman.

With a Groundwork of Educational Fundamentals, upon Which Experience Afloat May Build the Finished Naval Officer

The academy does not attempt to turn out finished naval officers. Without adequate experience afloat, as officers, it would be a hopeless task. This is an important point, and one often discussed in the Service. The Naval Academy turns the graduate over to the Service, for seasoning and further training in the school of experience.

That Healthy Minds in Healthy Bodies Are Necessities

It is recognized that human nature contains certain unfavorable elements as well as favorable ones, and that these must be combated by physical, as well as educational, means. The body must be healthy, if the mind is to be so. The American youth is frequently equipped with a super-abundance of energy, which must be diverted into proper channels, to prevent it from flowing into mischievous—or worse—ones.

Athletics thus serve a number of purposes. Through personal association, they satisfy the gregarious instinct which demands company— that is, they prevent loneliness. Through the successes of our teams, they increase the midshipmen's pride in the Academy. They increase company, battalion, class, and Academy spirit. They fit the graduates to coach ship's athletic teams. And what is more important, the general athletic system, completely standardized, does much to keep the regiment healthy, and builds up their physiques to keep pace with their constantly developing minds. It develops also certain qualities of leadership, such as good sportsmanship, instant decision, concentration, willingness to work hard, and self denial.

If the Graduates Carry into the Service Respect and Admiration for this Academy

The last part of the mission opens up a wide field of opportunities for good. The nature of the life is such that there will always be a feeling of relief when the coveted diploma has at last been won. Whatever is worth having, is worth working for; but there are so many slips between the cup and the lip, that success naturally means a release from an enormous strain—hence the spirit of "Thank God we're out of the wilderness." The feeling is perfectly natural, and there is no desire to change it. The aim is rather to have that "out of the wilderness" paean mean nothing more than the lifting of the safety valve, before the graduates settle down to a life of steady steaming, with heart and mind gratefully reminiscent of Academy days.

The policy is to give the midshipmen all the authority they are capable of handling. But this authority goes hand in hand with responsibility. They are held strictly accountable for their actions. The policy is to lead, rather than to drive. They are encouraged along a road, rough going at its best, and those who fail to measure up are summarily dealt with. A year's trial of this plan is encouraging but it will require one, or perhaps two, more years to establish definitely that the system is an improvement. The system is on trial, but gives every indication of success. The regiment realizes that the effort to relieve the grayness of Academy life means that, more than ever, they must measure up or take the consequences. There has been no let-down in discipline—quite the contrary.

Graduates who love the Academy will love the Service, and will make the necessary sacrifices, when the time comes, with a smile. They will come back to the Academy as Officer-Instructors, willingly, cheerfully, even eagerly. They will teach others to love the Academy, and in not so long a time there may be a waiting list for academy duty. Officers will be fighting for the detail. Then, as now, the midshipmen will be happy, interested, thoughtful, but with a greater sympathy for the Academy and the Service. This may come with the millennium; but at least we have a target.

3 *Discussion* of "The Naval Academy of Today and Its Mission"

Editorial from the *Baltimore Sun*,
31 July, 1923

EDITOR'S NOTE: *This piece originally appeared as an editorial in the* Baltimore Sun *on 31 July 1923, commenting on the previous article in this anthology. It was then reprinted in the "Discussion" section of the November 1923* Proceedings.

(SEE PAGE 453, MARCH, 1923, *PROCEEDINGS*)
An Editorial from the *Baltimore Sun*, 31 July, 1923.

A REGENERATING MISSION—During June, July and August the Naval Academy at Annapolis is engaged in one of the quieter phases of the general work to which its officials devote themselves, in one way or another, from year's end to year's end. When in June the graduating class has received its honors and assignments to duty, and the three other classes are on the regular summer cruise learning to apply the principles of seamanship which they have been studying, a new plebe class starts life at the Academy. It is a wise arrangement that allows these youthful recruits during the quiet summer season an opportunity to grow familiar with their new environment, to become accustomed to naval requirements

and imbued with the spirit of the great institution of which they are to
be a part. It is, perhaps, the most influential period of the four years,
because their minds are then in a peculiarly receptive state, and they are
likely to get in these introductory months impressions that are permanent, and to develop an attitude toward life far more important than
book learning or the mastery of scientific technicalities.

Maryland is particularly interested in the Naval Academy because
through its location at Annapolis it has become almost like the offspring
of her own heart and brain. But it is almost as close to the interest of
thousands of people in every other part of the country, and as the nurs-
ery of the Navy bears an intimate relation to national welfare as well as
to Patriotic pride. When a new plebe class, therefore, is flocking to the
colors and receiving its naval baptism something is going on that touches
deeply national fundamentals and national stability. A new naval legion
is in the making, and in the apparent calm of the Academy is receiving a
vital and lasting direction.

It is a rather trying time for the lads who have come together from
every section of the Union, some of them thousands of miles from home,
and unless wisely handled, some of them may become disillusioned
before they have fairly started, or get an entirely false conception of what
they have undertaken to do. For this reason the present superintendent,
Admiral Wilson, has been endeavoring to emphasize more and more the
moral purpose of our splendid naval training school. And we assume
that while the older classes are away getting acquainted with the element
on which they are to spend a large part of their lives, the new contingent
is having impressed on it the things which are above all else essential to
success.

What these things are, what the fundamental code of the Academy
is, was excellently stated in a paper by Lieutenant Commander Mahlon
S. Tisdale, United States Navy, in the issue of the United States Naval
Institute of March, 1923, entitled "The Naval Academy of Today and Its
Mission." Most people, and probably most of the boys who go to Annap-
olis with the title "Admiral" in the back of their heads, have supposed the

mission of the Academy was to train scientific fighters, to develop John Paul Joneses, whose souls would know no such word as surrender. That's some of it, but not all or most of it, as Lieutenant Commander Tisdale points out. The supreme mission of the Academy, as he defines it, is "to mold the material received into educated gentlemen, thoroughly indoc-trinated with honor, uprightness and truth, with practical rather than academic minds, with thorough loyalty to country, with a groundwork of fundamentals upon which experience afloat may build the finished naval officer, capable of upholding, whenever and wherever may be necessary, the honor of the United States." With this there must be also "healthy minds in healthy bodies," and "a humane yet firm and just dis-cipline" that will beget the "fullest efficiency" and a lasting "respect and admiration for the Academy."

With this fine text Lieutenant-Commander Tisdale elaborates his theme in a way which should be an inspiration to every midshipman who enters the Academy, and which gives the general public a new idea of what it is trying to do. "Character building," not sailing ships or firing guns, is its highest purpose. "High standards of character," said the pres-ent superintendent in a letter to the Secretary of the Navy last year, "are vital in a military organization. Men in other walks of life may trifle with the truth in everyday affairs and suffer little as a consequence. We cannot be so tolerant. The fundamental of an officer's value to the service is his trustworthiness."

Lieutenant-Commander Tisdale's pamphlet should be a moral text-book for every class at the Academy. Indeed, it might be studied with advantage by civilians in general and other educational institutions in particular. For at a period when honor, truth, duty, responsibility and unselfish patriotism seem to be losing their force and meaning in other spheres, this voicing of the mission of the Academy sounds a stirring trumpet call to higher ideals, not only to the ardent youth within its gates, but to every American man and woman. Only through regenerat-ing influences such as these can the country find the road to moral health and safety.

"The Naval Academy as an Undergraduate College"

4

Senior Professor Earl Wentworth Thomson,
U.S. Naval Academy

U.S. Naval Institute *Proceedings*
(March 1948): 271–85

"SAY, PROF, WHAT'S THIS HOT DOPE about the midshipmen strolling to and from classes at Annapolis? . . . What's happening at the old Navy school?"

The questioner was a Commander, graduated in the mid-thirties from the Naval Academy; the one being questioned was myself, Senior Professor of the Department of Electrical Engineering. The time was July, 1947; and the place, Northwestern University where both of us were in attendance at the Instructors' Course of the Naval ROTC. I assured this graduate that the rumor was only partially correct, that only the first classmen were allowed the privilege of not marching in formation, and that the standards and spirit at Annapolis had not been lowered, but had improved in the past decade.

The Naval Academy, being a public institution, and having as student representatives from 48 states, is under constant criticism. Some of this is correct and constructive, but much of it is based upon inaccuracies of data, warped emphasis on existing facts, and misconceptions as to the purpose and function of the training. Even among our own graduates we find personal experiences outweighing statistical tabulations, and the patina of time increasing emphasis on anomalies and fading out the normal method of behavior. We still find the sophomoric tales of the "juice prof who gave me a 1.0 when I rated a cold 4.0," the time the "cops and robbers papped me for insolence, when all I was doing was explaining," and "he's nothin' but a referee between the textbook and the red book."

The Naval Academy has a duality of purpose: first, to teach the fundamentals of education on the college level, and secondly, to point this education toward that required of junior officers. We cannot graduate the "compleat sub-lieutenant," but we can graduate men who have the intellectual potentialities of "officers and gentlemen," and are capable of a "future leading to the highest responsibilities of command, administration and policy."

The Superintendent, Rear Admiral Holloway, in a recent speech emphasized that the Naval Academy is the undergraduate portion of the Navy university system:

> I present the Naval Academy as the "college" of the Navy "university"—in other words the Naval Academy is the undergraduate level; the Postgraduate School and our General Line Schools, Staff Colleges, etc., operate at the graduate level; and we have numerous research activities, such as Bellevue, White Oak, Point Mugu, Engineering Experiment Station, Annapolis, and so on; all of which rolled up is more than equivalent to any ten universities operating at the collegiate, graduate, and research levels.

The Naval Academy must therefore be considered as a government institution on the college level, having a specific objective and attaining this objective through both practical and theoretical instruction.

Under the Post-War Plan for the Procurement and Education of Naval Officers the Naval Academy has been retained as a 4-year undergraduate school, the present academic year marking the complete return to the 4-year curriculum. The Academy, in cooperation with the 52 colleges and universities of the Naval ROTC's, will be the source of supply of junior officers for the post-war Navy. As such its present high standards of education must be maintained.

In 1802 West Point opened with ten cadets and became America's first technical institute, as for years it was the only school of applied science to give systematic instruction in those branches of learning which now are said to constitute "engineering." The second technical school was founded in Troy, N.Y., in 1824 as Rensselaer School. Among the firsts in education at Rensselaer were individual laboratory work, field experimentation, observation trips, scientific research on the graduate level, and the break with the traditional classical curriculum. Union College and the Naval Academy followed in 1845, with Lawrence School at Harvard in 1846, and Sheffield School at Yale in 1847. As late as 1874 West Point and Annapolis were still considered the leading engineering schools, or as one 1878 graduate recently reported: "The engineering course given at the Naval Academy was excellent, advanced, and thorough for its day, although little teaching was done and in my day I only heard two lectures." The Naval Academy since that time has broadened not only its curriculum but also its methods of instruction.

Since my return from the Pacific in November, 1945, I have visited, under orders from the Superintendent, numerous institutions comparable to the Naval Academy "to investigate the courses in chemistry, physics, and electrical engineering with a view toward bettering those at the Naval Academy." These institutions have included Massachusetts Institute of Technology, Worcester and Rensselaer Polytechnic Institutes, the United States Military and Coast Guard Academies, North Carolina State, Georgia Tech, the University of Illinois, and Purdue. The required curriculum and methods of education at these schools are not all the same, nor are they the same as at the Naval Academy. We could well adopt some of the methods of the civilian institutions, if the necessary time in hours per week, personnel, and equipment were available. In turn there are many ideas at the Naval Academy which could well be adopted by the civilian institutions. They can learn something from us, we can learn something from them.

During the war the Military and Naval Academies have been in the forefront in training leaders to win the war. It is a paradox of democracy that now that the war has been won these institutions which so successfully trained the leaders are subject to criticism. Let us hope that this criticism is based upon correct data. If you will read the succeeding pages you will find the picture of the Naval Academy as an educational institution in the academic year 1947–48.

Lectures on Education and Methods of Instruction

As an indication of the increased interest in education at the Naval Academy there was held during September a series of lectures conducted by Dr. A. John Bartky, Dean of the School of Education at Stanford University. The subjects discussed were "The College Curriculum," "The College Student," "Methods of Teaching Academic and Laboratory Subjects," "Tests and Measurements," and "Teaching Leadership." As part of each lecture there was formed a discussion panel in which were heads of departments, officer instructors, and members of the civilian faculty. The discussions that resulted showed a high degree of objectiveness, and showed that the Academy faculty believes thoroughly in self-evaluation and self-analysis. Dr. Bartky found that an argument could be started on nearly every phase of Naval Academy methods, and that, like all college faculties, there was no such thing as unanimity of opinion.

The Naval Academy uses all the methods of instruction found in civilian institutions, but not in the same degree or with the same emphasis. The major method still remains the classroom recitation preceded by study by the midshipmen. The classroom time will be used by the instructor for instruction and testing, for informal lecturing, seminar, discussion, oral questioning, explanation of difficult and important points, work at the boards, and written quizzes. The returning graduate will find a greater percentage of time being spent in instruction and not so much on quizzing. In the engineering and professional departments there will be found laboratories, where the students are working as individuals, in pairs, or in groups with each man having a specific job. In the professional and executive departments you will also find drills, for the purpose of discipline or of learning-to-do by actual operation. The coach-pupil method is found on the rifle range and in plebe drawing. Formal lectures are held in many departments: in physics and chemistry for demonstration purposes, in the humanities to interest the student in world affairs. Training aids will be found wherever applicable: charts and maps, sound-films and slide-films, models and mock-ups. In spite of many changes the classroom still remains the backbone of the Naval Academy system, but as stated previously the time spent in teaching has increased, that spent in testing has decreased.

Time Is of the Essence

The $64 question at the Naval Academy at present is how to get more hours in the day and in the week. The Electrical Engineering Department wants more time for physics and laboratory work, English wants more emphasis on the social-humanities, Executive wants more for infantry drills, parades, and the course in leadership. The officers of the fleet want more professional subjects, the PG's want more intense study in the fundamentals of science, the white-collar attaché crowd wants more "diplomatics" and languages.

At present practically all the heads of departments, executive officers, and senior professors seem to be working on new schedules and changes in curriculum. However, the shift has been made from the 3-year, 4-term, to the 4-year, 2-semester schedule. In March, 1943, the Superintendent, Rear Admiral Beardall, appointed a committee on postwar curriculum. This committee and its successors have been faced with the difficult problem of adding subject matter to the curriculum—aviation, more social-humanistics, leadership, electronics, and nuclear reactions—without increasing the midshipman work-load. Increases in time were given to Aviation, English, History, and Government, and to a lesser extent Executive, with the major losses being borne by the professional departments, Seamanship and Navigation, Ordnance and Gunnery, and to a minor extent by Marine and Electrical Engineering. Last year the 4-year schedule called for 3443 "contact hours," and, including study time, a midshipman work-load of 43.0 hours per week. The 1947 curriculum committee felt that this work-load was too heavy and cut 151 "contact hours" off the curriculum, making the average weekly work-load 41.7 hours.

Like Mark Twain's weather, everybody talks about the curriculum and yet no one department is completely satisfied with its place on the curriculum or the schedule. The healthy sign, however, is that, unlike the weather, something is being done about the curriculum, work-load, and scheduling. These are being continuously studied, and some solution will certainly be adopted. If all the changes that have been suggested by our well-wishers and our critics were adopted, the midshipmen would all be psychoneurotic cases, from their frustration attendant upon trying to carry the increased load and follow an impossible schedule.

Entrance Requirements

One of the matters which is being studied continuously at the Academy is that of entrance requirements and examinations—whether these are such as to fit normally into the American educational system, and whether the best screening is being accomplished. There are at present three methods of qualifying for admission at the Naval Academy. The most popular of these is the "College Certificate Method" by which 43% of the midshipmen have entered in the past ten years. Next is that of "Regular Examinations" which represents 35%, and lastly, the "Certificate with Substantiating Examinations" which admits 22%. One test of the validity of any one method in comparison with the others would be the per cent of academic failures under each method, but this shows little difference: 17% of those entering by the college method failed, 15% of those entering by the certificate-substantiating method, and 18% of those entering with regular examinations.

The history of the changes in entrance examinations is a long one, but we shall go no further back than World War I. The classes of 1922 to 1928, which had been examined for entrance from 1918 to 1924, showed an average over-all attrition of 40.5%. This was considered much too high. During this time the high school certificate was accepted without examination. This gave a poor screening, because of the wide variation in secondary school standards throughout the nation. Substantiating examinations were therefore considered necessary and were adopted in 1925. In 1935 the college certificate method was adopted so that men would not have to leave college in order to prepare for the Academy examinations. By this method a year's work in a college, university, or technical school accredited by the Naval Academy is considered the equivalent of an examination.

During the last year conferences on the subject of our entrance requirements were held with representatives of the National Association of Secondary School Principals, the National Science Teachers' Association, and the U.S. Office of Education. It was decided to drop Chemistry from the entrance requirements and regular examinations, to take effect in the April, 1948, examinations. The regular examinations will now be in English, U.S. History, Physics, Plane Geometry and Plane Trigonometry, and Algebra, with the substantiating examinations in comprehensive English and Mathematics.

In 1941, by Act of Congress, the age of admission was raised from between 16 and 20 to between 17 and 21, as of April 1. The average age of admission has varied but little in the past thirty years, as it was 18 years, 7 months in 1918, and 19 years, 0 months in 1946. It is the policy of the Superintendent and the Navy Department to bring the average entrance age down to the lower part of the band set by law. Not only do the records show that the younger men are much better adaptable for indoctrination into the Navy "way-of-life," but they also actually do better academically coming directly from high school than via the colleges. During the past year a study has shown that the top hundred men in scholastic standing of several classes were 4.3 months younger on the average than the bottom hundred. It is therefore considered advantageous to seek methods to lower the average age of entrance without lowering the standards.

Two factors have been instrumental in negating this effort and in raising the mean age of entrance: first, the assumed need for extra preparation of those taking the regular and substantiating examinations, and secondly, the college certificate method. Parents whose sons have been fortunate enough to secure nominations for appointment to the Naval Academy feel that it is their duty to leave no stone unturned to insure that their sons pass the examinations. Hence as much special preparation as they can afford is provided for the candidates. The college certificate method, although it adds a year to a man's age, allows the candidate to avoid examinations and gives him a year of college credit if he fails to get into the Naval Academy.

In the past ten years 70% of those who entered the Naval Academy have had college, preparatory school, or special preparation, leaving only 30% of the plebes who have come directly from high school. A five-year study shows that 33% of those who took the substantiating examinations passed them; but that, of those passing, 95% had had special preparation. Forty per cent of those who took the substantiating examinations passed; and of these, 75% had had special preparation. In contrast are the state universities which in many cases must accept any high school graduate from their state schools who has the proper credits. No wonder their attrition in freshman and sophomore years is excessive.

The Naval Academy, without actually making the statement, has always been adverse to the methods of "cramming" schools, and has tried to accept the best products of the legitimate public and private preparatory and secondary schools directly. All present changes in entrance requirements are beamed toward that idea.

Commencing in 1946, the College Entrance Examination Board, with its more than forty years' experience in testing methods throughout the forty-eight states, was given the task of preparing and grading our entrance examinations. With the exception of the theme part of the English examination, which is corrected at the Naval Academy, the examinations are entirely of the objective type. One of the recent developments has been the construction of a Naval Academy Aptitude Test which is an improved version of the Officer Qualification Tests used so successfully during the war to screen officer candidates and also used in the 1946 and 1947 Naval Academy entrance examinations. The aptitude test has been lengthened from one to three hours, including one hour of mathematics, and one-half hour each of verbal, common-sense science, spatial relations, and non-verbal tests. Additional to the candidates taking the regular or substantiating examinations, those certifying by the college method are required to take the aptitude test. There is no passing or failing, and no candidate is disqualified by his score on this test alone. It is used mainly in the evaluation of a candidate's ability and more particularly to assist in the examination of a candidate's certificate. With more experience in the study of performance curves and prediction grades, it is expected that greater use can be made of the aptitude type of testing which may lead to requiring a passing mark.

This Navy-College Aptitude Test was given in the summer of 1947 to 600 plebes who earned a mean score of 138. The mean score of *successful* NROTC candidates screened nationally for entry into the freshman classes of the 52 NROTC's in the fall of 1947 was 137. This rather amazing parallelism of mean score indicates proper screening, as the successful NROTC candidates were scholastically from the top 10% of those who took the test. In other words, the Naval Academy plebes and the NROTC students represent the top 10% of the nation's freshmen.

The screening of candidates at the Naval Academy through the system of appointments, entrance requirements, and examinations can be considered a success, particularly in view of the low attrition of the past few years. However, any change in method which will secure better students and get a higher percentage directly from high school would be considered an advance.

Appointments

The Superintendent in a recent memorandum to the Secretary of the Navy stressed the advantages of the present system of Congressional appointments. He stated that "one of the best indications for diagnosing future success is to look at the parents and background." Each Congressman assists in securing superior candidates for the Naval Academy because he can make a careful personal appraisal of those subjective attributes of character and leadership which it is so difficult to measure by an objective examination.

If all the appointments to the Naval Academy were filled, and the attrition were zero, there would be at the Academy this year a total of 3,897 midshipmen instead of 2,880, with an increase to 4,406 in 1950–51. Each Congressman and Senator is allowed five midshipmen at the Academy at one time, this accounting for 2,665. Also the President can appoint at large a total of 75 each year, with an additional 160 each year coming from the enlisted men and 160 each year from the Naval Reserve. The Naval ROTC and honor schools are allowed 20 each year, a total of 26 can enter from Puerto Rico and the American Republics, and 55 can come from other sources.

It is not expected that the total number of midshipmen will approach the 4,406 mark allowed under present legislation, but the increase in the Presidential, enlisted, and Reserve quotas means that the Naval Academy must expand in order to accommodate somewhere near 3,500 midshipmen in September of each year.

Psychometric Tests

From 1930 to 1947, with the exception of 1932 and 1933, there were administered to the plebes of the entering class in September the psychometric tests of the College Entrance Examination Board. These consisted of tests on scholastic aptitude, mathematics attainment, and spatial relations. During this time constant studies were made by the CEEB and the Secretary of the Academic Board as to the validity of these tests in predicting success or failure in the Naval Academy. For example in the Class of 1945, at the end of the first term of plebe year, out of 54 bilgers 48 had previously been predicted as poor risks. At graduation 72% of the predicted poor risks had been found deficient, and had bilged or been turned back. For the class of 1946, 90% of the predicted poor risks bilged.

These tests have shown particularly that turn-backs are poor risks. The class of 1945 included 50 men turned back from senior classes; of these 24 were considered poor risks, and 22 obliged the "statistics man" by actually bilging. Because no single test could be considered 100% accurate, and because legislation required other tests, the Naval Academy was unable to adopt the results of these tests to screen unsuitable midshipmen. However, the Naval Academy Aptitude Tests should accomplish this purpose in a similar manner.

Attrition

In the sixteen classes at the Naval Academy from 1932 to 1947 there have been 12,182 midshipmen (including turn-backs), of which 8,941, or 73.6%, have graduated with the B.S. degree, and of which 8,318 or 68.5% have been commissioned directly upon graduation into the Navy or the Marine Corps. The average attrition over this period has been 26.4%, with the class of 1940 holding the dubious record of 41.0%, and 1945 being given the prize with only 18.2% attrition.

Over the same period the average attrition during plebe year has been 14.6%, during youngster year 8.7%, during second class year 3.1%, and during first class year 2.2%. During plebe year the attrition from all causes has varied from 27.7% for the class of 1923 in 1919–20, the first full academic year after World War I, to 6.5% for the class of 1948, in 1944–45, the last year of World War II.

ATTRITION: 1932–1947		
		Percentage
Total midshipmen, including turn-backs	12,182	100.0%
Graduated with B.S. degree	8,941	73.6%
Total attrition	3,241	26.4%
Academic deficiency	1,825	14.9%
Physical disability	457	3.7%
Turn-back to lower classes	464	3.8%
Voluntary resignations	284	2.3%
Dismissed for bad conduct and other reasons	149	1.2%
Died	26	0.2%
Miscellaneous	36	0.3%
Commissioned at graduation	8,318	68.5%
Graduated, but not commissioned	623	5.1%
Discharged for physical disability	343	2.8%
Voluntary resignations	268	2.2%
Discharged for inaptitude	12	0.1%

ATTRITION BY CLASSES			
Class	**Membership, Including Turn-Backs**	**Attrition**	**Per cent Attrition**
1932	623	202	32.4%
1933	634	202	31.9%
1934	662	199	30.0%
1935	610	168	27.0%
1936	350	88	24.9%
1937	446	123	27.6%
1938	612	174	28.4%
1939	879	298	33.9%
1940	773	317	41.0%
1941	602	203	33.7%
1942	781	218	27.9%
1943	811	196	24.2%
1944	982	216	22.0%
1945	1110	196	18.2%
1946	1289	243	18.7%
1947	1018	198	19.4%

In a tabulation on attrition similar to the above, one never knows what to do with turn-backs. Like the poor disembodied souls floating in purgatory, they are neither in heaven nor have they bilged to hell. In the above table turn-backs from senior classes have been included in the gross membership of the class, and turn-backs into junior classes have been included as part of the attrition.

Several conclusions can be drawn from the above tabulation of attrition: (1) academic deficiency still remains the greatest single cause for failure, representing over half of the total losses; (2) physical disability is a poor second, with 3.7% before graduation, and an additional 2.8% who cannot be commissioned; (3) the total attrition at the Naval Academy is less than that in equivalent civilian institutions; (4) the attrition decreased greatly during the war years, reaching a minimum with the class of 1945 which graduated in June, 1944. Many factors have been credited with this decrease: better entrance examinations, closer physical examinations, better instruction, and a greater motivation on the part of the individual because of the immediate demand for more officers. It is believed that none of these is the whole story. With the return of peace, and with a raising of standards, even with a better screening before admission it is expected that the attrition will increase. This has always been the pattern of the past.

Bachelor of Science Degree and Accreditation

In the spring of 1933 authority was granted by Congress to the Superintendent of the Naval Academy to confer the degree of Bachelor of Science upon the graduates of the Academy. At the same time the Military and Coast Guard Academies were given the same privilege. Granting of the degree was made contingent upon the accrediting of each academy by the Association of American Universities, but this accreditation had been accorded the Naval Academy in 1930. This Association approves undergraduate schools so that the graduates may enter any one of 33 member universities for graduate study.

The endeavor to secure legislation allowing the granting of the B.S. degree had started in 1922 when a bill had been prepared but had failed of passage. When in 1933 only 58% of that class were to be given commissions in the Navy, it was believed that a degree would be of great help to the graduates who were to go into civilian life. In 1937 a further bill was passed allowing all previous graduates to be given the degree. The granting of the degree was an excellent move, as it gave the graduate of the Academy a standing on the same level as the graduates of the colleges, and allowed him to take graduate work in universities leading to advanced degrees without a course-by-course analysis of his record.

Some of the older graduates chuckled with inward satisfaction when they received the Naval Academy B.S. degree after acquiring many higher degrees elsewhere. The late Dean Emeritus Cooley of the School of Engineering of the University of Michigan received his Naval Academy B.S. degree 60 years after his graduation in 1878, and ten years after he had retired from active educational work and from the presidency of several scientific and engineering societies.

Many Naval Academy graduates falsely believe that this B.S. degree is a B.S. in Electrical Engineering, or General Engineering, or even in Naval Science, but this is not the case. The degree is a straight B.S., with the major field undesignated, and with the meaning of the degree unlimited.

In order to take advantage of greater opportunities through liaison with other educational institutions, the Naval Academy has recently joined several of the cooperative educational societies. In 1941 the Academy was granted institutional membership in the American Council on Education, and in January, 1947, was voted membership in the Association of American Colleges. In the summer of 1947, after investigation of the curriculum and the methods of teaching, the Academy was accredited by the Middle States Association of Colleges and Secondary Schools. This is the chief college accrediting agency in this area, having its counterparts throughout the country in the New England Association, Southern Association, North Central Association, and Northwest Association.

Through the legislation granting the degree of B.S. and through the accreditation by educational societies, the Naval Academy has been recognized as an excellent educational institution at the undergraduate level. This has been particulary necessary since the expansion of the NROTC among 52 degree-granting colleges. The Academy graduate must be equivalent to the ROTC graduate, even on the degree standard established by those institutions. There has never been any question about the standards of the Academy, it has only been necessary to get these standards recognized by our collateral educators. We should not stand alone in the admiration of our own glory.

Rules, Rates, and Regulations

The present administration is continuing the determined effort to obtain the optimum synthesis and understanding of Leadership and Discipline. The volume *Naval Academy Regulations* still remains the most powerful "essence of good and evil" for the midshipmen, but a definite trend is indicated by recent changes. The first classmen are being given more responsibility for leadership in the brigade, and as they show that they rate them are extended privileges "commensurate with their responsibilities." This has tended to increase the midshipman's individual initiative and sense of leadership. The first classmen do not march in formation to and from classes, but walk in chatting groups as if they were officers at one of the graduate or staff colleges. All the other classes still march in formation, except upon dismissal from the last class in the afternoon. For a time "route step" was tried, but the result looked like Dryden's "rude militia" and was discontinued.

Each first classman, with certain provisos as to conduct, aptitude, and satisfactory academics, is allowed every other weekend to go away from the Academy. The second classmen have been given one weekend each term, as each desires. This privilege of weekends is guarded jealously by all, and in spite of numerous temptations has resulted in practically no increase in Class A offenses. Town liberty is also granted to first classmen every afternoon from the end of the last drill period to supper formation, and in addition they are now allowed to ride in automobiles.

Last year the system was tried of not assessing the first class with demerits for Class B offenses, but of requiring only a statement explaining the reasons for the offense. This experiment was not considered a success for several reasons: (1) because there no longer existed a yardstick for measuring the difference between the "good" and the "poor" from the viewpoint of relative conduct; (2) unnecessary time was being used up in writing the wrong kind of reports, when too often the culprit was "guilty as charged"; and (3) a few "sharpshooters," when desire overcame responsibility, were taking advantage of the relaxation of discipline. After a careful study of the advantages and disadvantages of the system the Superintendent made the decision to return to the demerit system, largely because good discipline and conduct were considered a necessary part of leadership and responsibility. Punishment, however, does not entail walking extra duty, but, parallel to that assessed an officer, includes restriction, confinement, and deprivation of privileges.

That many midshipmen are able to obey all the rules and regulations is probably unbelievable. However, in the past academic year 150 out of the 2,576 midshipmen had 4.0 in conduct as a result of collecting *no demerits*!

One of the traditions most firmly established at the Naval Academy has been that of "rates." These have been designed for indoctrination, on the principle that there should be "appropriate gradations of responsibilities and privileges by their seniority." The class of 1948-B, when it assumed the leadership of the brigade in the spring of 1947, published a "Class Policy" in which these rates by classes were re-established. The essential underlying philosophy was that there must be a "class attitude of brigade leadership based upon mutual respect, precept, and example," and that in correction there must be eliminated the "flagrant violations of mature dignity."

Many of the long established rates are still maintained—squaring corners by the plebes, the seating arrangement in the mess, the rates as to walks and ladders, snapping to attention and "sounding-off." But many of the old customs have been abrogated, such as the practices of "spooning," "sitting on infinity," stoop-falls, and impromptu plebe performances. The emphasis has been placed on "private man to man correction which will supplant all practices of hazing," and the assumption of leadership by the first class with its authority to correct. Plebes are still plebes, with few privileges to "Carry on" except after victories or excellent performance, but the specter of hazing and unofficial "running" is disappearing. Plebes are still questioned, but by order only, on "customs and traditions of the Navy, on professional knowledge, and on pertinent current events." Gone are the days of "Mr. Speaker, Mr. Speaker!" "Good Morning, Merry Sunshine," and the drills on "Man Overboard" and "Swimming to Baltimore."

This Class Policy, including the rates, has been approved by the Superintendent and given the "force of regulations and approved policy in all appropriate circumstances." The Superintendent in a recent "Open Letter to the First Class" emphasized the "offense of unauthorized assumption of authority" and declared that the "powers of command must be exercised only by those having professional competence, moral integrity, and instant, constructive, and open-hearted obedience to constituted authority."

The Executive Department is backing up the first class in the maintenance of this Class Policy with its rates. The report via unofficial channels from the brigade is that hazing is extinct, as the real powers in the first class are officially frowning upon the few sophomore sadists who would practice the "science" under the cover of darkness.

Marks and Marking

The late, unlamented dictator Hitler is said to have once remarked that he was not alarmed about the potentialities of a nation whose number one household god was a wooden dummy named Charlie McCarthy. By the same token the critics and detractors of the Naval Academy should not be exorcised because the idol of the Brigade is a wooden figure-head—Tecumseh, the god of 2.5. (Correction: the visible statue now is of bronze guarding the sections as they proceed to battle academics, but the wooden original still stands in Luce Hall—both Charlie and Tecumseh have had termite trouble.)

There has been in the past an overemphasis on marks at the Naval Academy, particularly in the traditional requirement for a daily mark in every subject. This necessity is changing in most of the departments, the emphasis now being rather on instruction. In "Juice," "Steam," and "Bull" the first and second classes are given a weekly mark, often the result of a single weekly quiz. The third and fourth classes are not treated so leniently, but the requirement for a daily mark has been abolished. A new Naval Academy standing order states that "it shall be within the discretion of each respective head of department to establish the procedure by which the weekly mark is determined in his department." This responsibility is being willingly assumed by the heads of departments. It has even happened that, because of material which required special instruction, the mark has been skipped for a week. Imagine that at the Naval Academy!

The objective now is to teach, with the testing and evaluation of midshipmen assuming a secondary though important role. This has allowed more freedom to the instructor in the classroom, as he is not now haunted by the necessity for a daily mark.

It must not be forgotten, however, that the testing of the student through examinations or quizzes is part of the learning process. Schools which have tried to abolish marks and examinations have consistently failed in this attempt. My own undergraduate school, Clark University, tried this in its early history, but the "*savoirs*" protested. They could not bear to be rated no better academically than "*Dummkopfs*." "Education for marks," however, is a poor substitute for learning.

The frequency and form of this testing is a much argued point with professional educators. In his *Manual of Mental and Physical Tests* the late Professor G. M. Whipple of the University of Michigan has stated: "Shall efficiency be measured in terms of quality, excellence, delicacy or accuracy of work, or in terms of quantity, rate, or speed of work?. . . No general answer can be given."

The Naval Academy has used many types of quizzes: subjective and objective, open and closed book, written work at the seats and at the boards ("Draw Slips and Man the Boards!"), reports, short "research" papers, homework problems in physics or phrases in languages. The choice is left to the head of committee, of detail, or to the individual instructor. The Department of Electrical Engineering has for years used objective type questions in the examinations: true or false, short answer, fill-in, matching, multiple choice. The swing at present is probably back to the subjective or essay type question, and to problems with mathematical answers. The chemistry examinations still include 60% objective questions, the physics 30%, and basic electrical engineering 20%.

The Naval Academy, partly because of the necessity for accurate grading, requires excellent preparation for all examinations. Each head of a class committee or detail prepares the examination from the important material covered. Average instructors then work it against time. In Mathematics the instructors must work a 180-minute examination in 60 minutes; in English they must work it in 75 minutes, and in Electrical Engineering in 90 minutes. The examination is then given a final polishing, and ambiguous and unimportant questions are discarded before the examination is approved as ready for the midshipmen.

From the above it is seen that the Naval Academy still regards marks as important, but that it is coming to believe that 700 valid marks have the same accuracy in determining a midshipman's final standing in his class and on the promotion list as the previously required 2,500.

Training Aids

During World War II the Army and Navy in their training camps and schools adopted the policy of maximum use of training aids, also known as visual aids and special devices. These aids were used at all levels of intelligence, from the training of illiterates to read and write to the daily briefing in the War Room of the Combined Chiefs of Staff. These aids included sound and silent moving picture films (both 35 mm. and 16 mm.), slide-films with accompanying records, charts of lecture outlines or of the breakdown of weapons, models of every conceivable piece of equipment, mock-ups of materiel from a wooden Springfield to an expanded B-29, wire recordings for pronunciation and speech training, war games including models of ships and terrain, three-dimensional maps, bread-boards for fire control and electronic devices, and others too numerous to tabulate.

For the past few years there has been on duty at the Naval Academy a Training Aids Officer whose duties have included advising heads of departments on the use and procurement of training aids and special devices. In each department an instructor has also been assigned as training aids officer. A large library of sound- and slide-films, both from Naval and civilian sources, has been collected. Many of these are used regularly in the various departments for instruction in classrooms and at drills.

The main philosophy regarding training aids at the Naval Academy is that for instruction purposes the real thing is far superior to any training aid; any device except the real article is merely ersatz. The movie can never take the place of a real teacher, the mock-up can never completely substitute for the real weapon, the beautiful case of three-dimensional cotton cloud formations can never take the place of direct experience with towering cumulus clouds in an airplane. The model and the chart are functionally incorrect unless the student can do something—trace circuits, operate equipment, or ask questions, and not merely stand and gaze. The training aid must be used to assist, and not to substitute for the teaching process.

With this in view the laboratories of Electrical Engineering have real full-size motors, generators, oscilloscopes, waveguides, and electrical instruments; the laboratories of Marine Engineering have real engines together with working cutaways of these real engines; Ordnance and Gunnery has the actual 5-inch, 38-caliber double purpose guns with regular naval fire control equipment. To assist in understanding the real, operating range-keeper there is a breadboard model which works, together with numerous explanatory charts. However, as it would be rather impracticable to move an actual shore bombardment into the laboratory, this must be done with a simulated small scale model around which the midshipmen can work, each one having his position of responsibility in the problem. Radar can be taught from the textbook and the laboratory in Electrical Engineering, and from the actual instruments controlling the guns in Ordnance.

Wire recorders are being used in English for remedial speech training and for training in public addresses, and in Languages for pronunciation. The individual hears himself as others hear him, and sometimes this is quite a shock.

The conclusion is that training aids are valuable to assist in instruction, but should not be used to replace good teaching. The Naval Academy is conscious of this limitation and is proceeding in their use accordingly.

Members of Civilian Faculty

Probably the greatest single advance educationally in the Naval Academy during the past ten years has been the increase in the size of the civilian faculty. In 1938–39 there were 59 civilians and 188 officers attached to the academic departments, the civilians representing 24% of the instruction staff. In 1947–48 there are in these same departments 156 civilian and 222 officer instructors for a civilian ratio of 42%. The above figures do not include the civilian organist, the librarian, the Assistant Secretary of the Academic Board, the officers of the Executive Department, nor the officers and civilians of the Department of Physical Training. Including the civilians among these, and two professors on leave without pay, there are 176 members of the civilian faculty.

For the academic year 1947–48 the Superintendent succeeded in securing an original allotment of 209 members of the civilian faculty from the Navy Department, but this was cut to 185 by budget requirements, and to 176 by an across-the-board cut in government appropriations at the last moment. At the same time as this last cut the allowance of officers over the whole Severn River Naval Command was decreased by 116. The Naval Academy is still short 15 to 20 officers below this reduced minimum.

PERSONNEL OF DEPARTMENTS (Including Heads of Departments and Executive Officers)				
	1938–39		1947–48	
	Officers	*Civilians*	*Officers*	*Civilians*
Seamanship and Navigation	35	—	26	—
Ordnance and Gunnery	22	—	25	—
Marine Engineering	40	2	69	8
Aviation	—	—	19	—
Mathematics	12	24	9	59
Electrical Engineering	41	4	54	21
English, History, and Government	20	16	12	43
Foreign Languages	18	13	8	27
Total (Academic)	188	59	222	158
Executive	27	1	35	1
Physical Training	13	9	5	15
Asst. Sec. Acad. Bd. and Librarian	—	—	—	2
Total	228	69	262	176

Recommendations have been made that the faculties of the Departments of Mathematics, English, and Languages include 75% civilians, that Electrical Engineering increase to 50% and Marine Engineering to 25%. There should be little opposition to this increase in the non-professional departments, as the civilians provide the continuity of instruction and tenure of the academic staff.

The existence of the civilian professor has been made more secure at the Academy by the publication of a definite policy of pay, promotion, and tenure. The government is cooperating with the Teachers Annuity Association toward the purchase of deferred annuity policies to which both the professor and the government contribute 5% of the salary, this being collectible at age 65.

New appointees to the grade of Instructor now must have at least a master's degree (M.A. or M.S.) and have engaged in at least one year of teaching on the college level. It has become a policy that the Doctor's degree (Ph.D., D.Sc., etc.) will be required for promotion to the grade of Professor. This should not be too difficult as 38 of the faculty now have their Ph.D. degrees, with numerous others in the process of earning them.

For several years money has been available for civilian travel for the purpose of attending scientific and educational meetings and conventions, and for visiting other institutions of learning. Many of the faculty members, both officer and civilian, have participated in these meetings, and have helped secure a closer relationship between the Academy and other institutions.

In the five departments with civilian professors a new rank has been established, that of Senior Professor. In Languages this Professor is known as the Faculty Chairman, in Mathematics and English as the Dean, and in "Juice" and "Steam" merely as the Senior Professor. His duties are numerous and variable. He may act as chairman of a department advisory board, serve as liaison officer between the civilians and the head of department, act as advisor to the head of department, be the coordinator of curriculum, instructor assignments, teaching methods, examinations and scheduling, and the general factotum in all matters of continuity and past experience.

It is essential in an educational institution that there be professional educators, men who have a background of training well beyond that of the undergraduate level or the daily lesson. There exists no substitute for knowledge of subject matter for an instructor. The civilian members of the faculty supply this additional knowledge and continuity. Their place is necessary and must be maintained secure if the Naval Academy is to remain a top-drawer educational institution on the undergraduate level.

Midshipmen's Pay

On July 1, 1947, the midshipmen's pay was raised from $65.00 to $78.00 per month, but because of increasing costs the "take home" pay still remains an infinitesimal quantity. For example, laundry now costs each midshipman $101.00 per year, and a suit of uniform blues which ten years ago could be bought for $39.00 now is debited at $58.26. The mess ration has been increased to $1.20 per day, but as this is credited and debited at the same time on the midshipman's account, there can be no saving here. This ration includes the price of food, which has gone up so much faster than the ration value, the cost of mess gear and of any special diets for the athletic squads on the training tables. The midshipman has no housing problem, at least for himself, as each is given his "quarters" complete with bed, desk, closet, and shower.

Material Expansion

The Naval Academy, being a government supported institution, must depend upon the will of Congress for money to provide needed expansion. In time of peace the securing of this money is a long and tedious process. At the time of World War I, there was constructed for the Department of Marine Engineering another wing on Isherwood Hall, this being called Griffin Hall. In 1920 Seamanship and Navigation, with Foreign Languages, moved into an excellent building on the seaward side of the gymnasium, this being called Luce Hall. From then until 1937, a period of 17 years, there was no expansion in the academic buildings at the Academy. Peace was too promising. It is true that the Natatorium was built in 1924, the funds being so meager that the construction was of yellow brick rather than of grey granite; and that the Athletic Association gave the money in 1930 for a new boathouse on College Creek, now known as Hubbard Hall.

In 1937 another wing was added to Isherwood Hall, this being called Melville Hall. The Department of Marine Engineering uses the lower floor of this for an internal combustion engine laboratory, and the upper floor for a drawing room. The drawing rooms of Isherwood, Griffin, and Melville Halls can now accommodate a plebe class of 1,050 for drawing, examinations, or psychometric tests, being interconnected with a modern public address system.

In 1939 the Naval Academy Museum was built on Maryland Avenue, just inside the main gate, from funds provided by the Athletic Association and the Naval Institute, which have offices on the second deck. The Museum has expanded so greatly, with the addition of World War II memorabilia, that something will soon have to be done about expansion or discriminatory selection. In 1939 the yard dispensary was moved from its old quarters in Mahan Hall to a new dispensary behind the Museum, and in 1947 this building was revamped as an Aviation Building and the yard dispensary was moved to the Naval Hospital. Sick bay was moved in 1940 from over the rotunda of Bancroft Hall to the sixth wing where it now resides. Among the M.D. quarters only "Misery Hall" in the gymnasium has retained its position during the years.

For many years the Chapel was too small for seating the regiment of midshipmen. In 1939 it was found necessary to enlarge it by the addition of a wing, so that the main chapel now seats 2,347. A small chapel, called St. Andrew's, was included on the ground level of the new wing. As a large number of the brigade regularly worship in the churches of Annapolis, the present chapel, although crowded every Sunday, has a sufficient seating capacity for the present brigade.

In 1941 McDonough Hall, the gymnasium, was divided horizontally, to give two floors where one had existed before. The Department of Ordnance and Gunnery at the same time expanded into a new recitation building, named Ward Hall. This is connected with the armory by a passageway so that ordnance drills can be run simultaneously in the armory and lower floor of Ward Hall.

Even with the new academic buildings the total number of classrooms now available is only 151, which is not enough to handle the present 228 sections of the brigade. The first need in any expansion is for increased classroom space, if the smaller size of sections—from 12 to 15—is to be maintained.

Bancroft Hall has seen several recent changes, the main one being the addition of two new wings in 1941, each housing 310 midshipmen in double rooms. The normal housing capacity of the Hall is about 2,500, with a saturation point of 3,100, thus still making it the largest dormitory in the world. Earlier in 1934 a "steerage" was added on the level with Smoke Hall, this now being complete with soda fountain and hostess. The job of modernizing the four old wings has just been completed at a cost of $4,300,000. All the rooms have been revamped as double rooms, new showers have been added, and fluorescent lighting installed. In fact the whole job has been so complete that the interior of Bancroft Hall has practically been rebuilt. The "watertight integrity" of Bancroft Hall is now an accomplished fact, and even the showers on the fourth deck spout water.

The Academy purchased about eight acres from the waterfront of Annapolis, just behind Thompson Stadium, and this is now used for drills and intra-mural sports. The old Johnson Lumber Yard, which used to pour smoke into the football stands in the middle of every game, has been moved to West Street extended, much to the joy of all football fans. Another intra-mural sports field was constructed by making a 23 acre fill off Hospital Point in the Severn River just inside the Ritchie Bridge.

Fourteen apartment houses have been built, each capable of housing six families. A total of 300 additional families are housed in Quonset huts in two Homoja villages, constructed under war contracts.

Plans are constantly being made for expansion, to handle 3,600 or even 5,000 midshipmen, so that when Congress appropriates the money there will be no lag between the appropriation and the commencement of the construction.

Conclusion

Several other titles were suggested by readers of the draft copy of this article—"How Are Things in Annapolis?," "Something New Has Been Added," or "The Naval Academy Evaluates Itself." You have seen that there is something of all three ideas in the previous pages, but above all this has been a "Professor's Report" on the status of the Naval Academy as an educational institution. You have seen glimpses of some skeletons, answers to the criticism of some razzing, raving, and ranting critics, but essentially you have seen the evolutionary plans for the progress of naval instruction on the undergraduate or collegiate level.

The Naval Academy must expand materially and educationally. There must be an airfield. As the Superintendent stated recently: "The Naval Academy without an airfield has no more sense than a Naval Academy without a harbor." We must graduate men better prepared intellectually for service in life and in the fleet. Our instruction must be on a par with the best in the country.

Many fond memories from the past still remain to grace the Yard and Annapolis:

> Tecumseh guarding the approach to academic walk with his annual coat of many colors;
> The sarcophagus of John Paul Jones supported by symbolic dolphins in the crypt of the Chapel;

The "tree" for the unsats early each week, where "Tis unpleasant, sure, to see one's name in print";

The sound of midshipmen's hurrying feet echoing through the silent streets of Annapolis bare seconds before the fatal hour of midnight;

Band concerts to distract our wandering minds at 10 A.M. and 4 P.M. with the 120-beat when we dare march by en route to classes;

Movies on Sundays when even the "yard engines" rate being dragged;

Memorial Hall with its memories of Articles of the Navy, and "Don't Give Up the Ship"—pictures have been added—our friends who did not return—Ike Kidd from Pearl Harbor—Dan Callaghan and Norm Scott from the *San Francisco* off Guadalcanal—Pinky Swensen from Savo Island—Ted Chandler from Lingayen Gulf;

The Museum with its mementoes—Halsey's saddles, Mitscher's cap, the surrender table from the "Ol'Mo'";

Midshipmen still "riding the velvet" going into exams;

The new stadium still only a gleam in the eye of the Athletic Association;

Traffic still jamming the narrow colonial streets of the ancient city.

Yes, things are normal at Annapolis. Be assured that the old Navy school is still progressing, and very much alive.

Educated in physics and mathematics at Clark University and Dartmouth College, with graduate work at the University of Chicago, Johns Hopkins University, and Massachusetts Institute of Technology, **Professor Thomson** served as Captain in the U.S. Coast Artillery in World War I. As Colonel, General Staff Corps, U.S. Army, he saw service in World War II in England with the Anti-Aircraft Command, the VIII Bomber Command, and the Eighth Air Force as Flak Officer and Chief of Intelligence. After duty in 1944 in the Plans Section, Army Air Force Headquarters in Washington, he went to Pearl Harbor as Chief of the Flak Intelligence Section of the Joint Intelligence Center, Pacific Ocean Areas. For many years a Physics professor at the Naval Academy, Professor Thomson is affectionately known by thousands of midshipmen and naval officers as "Slip-Stick Willie."

5 "The Naval Academy in Transition"

Major John E. Williams, USMC

U.S. Naval Institute *Proceedings*
(January 1949): 67–71

WEEK AFTER WEEK, month after month, the word comes floating back from the Fleet: "We hear the plebes have really taken over Bancroft Hall"—"What's all this dope about first class rating every week-end, and driving automobiles on liberty?"—"I hear that all Youngsters have radios and the plebes rate dragging; what in the world is going on down there on the Severn?" The foregoing, like all scuttlebutt, has finally come full circle. It is all scuttlebutt, it has originated outside of the Naval Academy, and it's all coming back to you right now.

It is not surprising that you have these thoughts. After all, a good many wardroom bull sessions have as their subject the Naval Academy, its shortcomings and its limitations. And to many of the rumors there is, of course, some particle of truth. As officers, it is natural that you should be concerned about and interested in just what is being done in the training of prospective career officers, and this interest should be held by non-graduates as well as by graduates of the Academy. If you are like most officers in the Navy, you have had no contact with the Naval Academy since your day of graduation (or none at all), so you have little to go on in formulating your thoughts regarding the place except what you hear in bull sessions, and what you read in the popular periodicals.

Both are potentially dangerous sources. Neither is likely to be altogether factual, and both are likely to be colored and distorted for effect. The officer expounding about the Academy in the bull session is probably not being deliberately malicious, but his remarks can be just as damaging as though he were an avowed saboteur. Chances are that all this type of individual is thinking about is being impressive. He wants all and sundry, and particularly those junior to him and those who are not graduates, to know that in his day the Academy was a very tough place indeed. You will recognize this individual at once when he begins his remarks with the all-too-familiar phrase, "When I was a midshipman . . ." The same man would in all probability deny vigorously that what was good for his father, or for his grandfather, is necessarily good for him, but he follows that untenable train of thought without a qualm in discussing the Academy. But still this reminiscer, this champion of the "good old days," is not nearly so dangerous as the self-appointed authority who gets his peculiarly slanted observations into the public print. We're not all "from Missouri." A whole lot of us tend to accept the printed word unquestioningly, or almost so. That is the big advantage enjoyed by the article writer—in most cases his audience is psychologically prepared to believe him because his utterances, even though conceived in bias and brought forth in anger, have been published. So it is apparent that a lot of loose talk about the Academy is being indulged in.

Some of it indicates that the plebes are living a country-club existence, with the first class too occupied with their own pleasurable pursuits to take a hand; and again some of it indicates that the whole institution is vice-ridden, corrupt, and a haven for sadists, with lashings and other extreme tortures of common occurrence. The truth is, of course, not to be found at either extreme. It is true that the Academy has changed, but in what way?

This article will not attempt to cover every aspect of Academy life, but will concern itself, for the most part, with life inside Bancroft Hall, as it is in this area that the widest misunderstanding, and the deepest interest, seem to exist. There will be no attempt here to "white-wash," or to justify for the sake of expediency. The intention is to present an objective view of life at the Academy as it is actually being lived, particularly within the purview of the Executive Department.

There has been one basic conceptual change in recent years, the effects of which have extended into every aspect of life at the Academy. The change has had to do with that abstract and elusive thing known as "Discipline"—what it is, and how it is attained. In the past, as many of you will recall, the state of discipline obtaining seemed to be regarded as a function of the number of conduct reports turned in. The more "fraps" that were handed out, the tighter the "discipline," and consequently the more efficient the administration. That was a nice, easy way of life; you could accept it unthinkingly, could enter into it with spirit, and all was serene. Duty officers ran races with each other to see who could most load the conduct sheets, the midshipmen served their extra duty, the "status quo" was maintained, and the Academy seemed to turn out pretty good officers anyhow. But in that system, however much fun it might have been as a game of "cops and robbers," there was absolutely no regard for the midshipman as an individual. There was no effort expended in trying to see behind behavior and get at the roots of trouble. Actions were punished, relentlessly and in accordance with an unvarying code, but causes for the actions were blithely ignored. You remember the pattern: you were caught in a delinquency (perhaps by a D.O. peeping from behind a bush); in due course the "frap" appeared on a table outside the battalion office; you initialed it, and in a day or two you and "Miss Springfield" started keeping company. No attempt had been made at correction or at determining your motivation; you were no better a man after three or four hours of extra duty than you were before you got into the trouble, and no one seemed to care. In the vast majority of cases, you would never have had a talk with your company or battalion officer. Any evidence of personal interest in you as a human being on the part of the officers of the Executive Department was wholly lacking, or very nearly so.

Now, fortunately, all that is changed. Yes, the conduct report is still used and standards are still as exacting, but the approach is different. D.O.'s no longer lurk behind monuments to pounce on the unwary; they don't vie with each other for the "Frapper of the Week" title; in fact, the current group of D.O.'s strive to be so straightforward in the execution of their duties that scarcely one of them has a nickname indicative of any personal peculiarity. The passing of the "characters" from the officers' watch list has been officially recognized by the *Log*. The editors of that magazine have not had a good "D.O. story" to run this year! But midshipmen are still placed on the report, although not in wholesale lots in a competitive atmosphere. The watch-officers today have learned to recognize and appreciate the difference between honest shortcomings and deliberate violations, and they counsel and correct as much as they report for disciplinary action.

But a more important change is to be considered after the conduct report has been handed in and logged. It is not left outside the battalion office to be initialed; it is sent to the Company Office of the midshipman concerned, where the latter initials it, usually in the presence of the Company Officer. This brings the alleged offender face to face with his immediate commissioned superior in the chain of command, and the two talk the case over. The officer has at hand a complete catalogue of all the midshipman's previous offenses; he knows the man from day to day contact; together, they can soon get at the root of the trouble. If the purpose of the regulation is not clear to the midshipman, it can be readily explained; if he has any excuse it can be heard at once; the indicated corrective steps can be instituted. The midshipman will still get some extra duty if he has been on the wrong side of the line, hut at least it will not all have happened automatically and inevitably. He will know why he is serving that extra duty, will know what his Company Officer thinks of him, and what he expects of him in the future. The whole transaction is more personal, more mindful of human dignity, and hence is naturally more appealing to all parties concerned. And, not the least important, it provides opportunity, through counseling, of planting and cultivating the seeds of self-discipline in the officer-aspirant.

And where, you might well be asking, did that company office come from? You remember that the Battalion Officer and two or three Company Officers all sat down together in an awesome cubicle, the insides of which you might never have seen in broad daylight unless you were in Class "A" difficulty. For all practical purposes, two or three of those officers might just as well have been back at sea. Yes, you saw them at noon meal formation and you saw them, fleetingly, on watch, but with how many of them did you ever pass the time of day? To how many of them did you ever carry any questions or make any suggestions? Today the Company Officer has been given some status, and his duties have been made meaningful. He has been given an office (in the form of a vacated midshipman's room) on his own company's deck, with a desk, a few chairs, and the full responsibility of guiding, counseling, and developing that company of some 95 midshipmen. A lot of little things have resulted from this change, the sum total of which is most important. The midshipmen pick up their dining-out forms, their requisitions, etc., in this office without hiking down several decks to the Battalion Office; they turn them in there for signature and a fast return. They come in and out of that office so many times that they grow accustomed to exchanging greetings with their own officer, get to feel that they know him as a person and not just as an authority, and a feeling of naturalness permeates their relationships. Most Company Officers have a coffee mess for first classmen in their offices, where the gentlemen of the senior class can drop in for a cup when their duties permit, just like human beings the world over. You "old-grads" are probably thinking we have gone mad for sure, but what tenable argument can be made against such procedures? Remember how much more at ease you felt the first time you put in an appearance in the Captain's Cabin and he had you join him in a cigarette and a cup of coffee, instead of requiring you to stand at rigid and nervous attention? It works the same way here. Ideas are much more likely to flow freely in a natural atmosphere than in one charged with an awareness of a large senior-junior gap. And that's the aim of the Company Officer—to break down the barriers of formality, in situations where formality is not called for, in order to get on with the business of coming to know and understand the midshipmen.

You have probably heard some talk, too, of the recently established Leadership course which is being given to the midshipmen by the Academic Section of the Executive Department. The Company Officers are the instructors in this course, in addition to being administrators and duty officers in Bancroft Hall. At present, eighteen of the twenty-four Company Officers are engaged twelve hours a week each in teaching Leadership to the first and second classmen, and next year all twenty-four will be in the act. It is one of the best developments yet introduced, this business of having the Company Officers teach the Leadership course. The informal nature of the discussion type classes enables the officers to get to know intimately many more midshipmen, and vice versa. One definite result of this system has been that, perhaps for the first time, the midshipmen have come to recognize the officers of the Executive Department as people—as brother officers—with problems and thought processes not unlike their own. And this constant interchange between the Company Officers, as administrators and instructors, and the midshipmen has resulted in an awareness of leadership on the part of both that was never before approached, even remotely.

That, in brief summary, covers the effect of the conceptual change in "discipline" at the officer level. How, you might ask, is it working out at the midshipman level, in their inter-class relationships? The answer to that question, to be fully appreciated, needs to be seen—to be experienced—rather than read, but perhaps a few words can give the general outline.

To understand how the present inter-class relationships evolved you will have to go back a few years to the Summer of 1944, when the recent first class (1948-B) entered the Academy. This, of course, was the mid-war period, when the attitude of "anything goes" was in full sway. From all accounts, '48 did have a rough plebe indoctrination. They told you this themselves, but they didn't do so in a spirit of braggadocio. They would also tell you that the flagrant cases were not general throughout the Brigade, as you might believe from reading the magazines. There were enough of these flagrant cases, however, to awaken a determination in those plebes that when they became responsible for running the Brigade things would be different. To this end, as second classmen, the men of '48-B got together and formulated an admirable instrument known as the "Class Policy of 1948-B," which was approved by the Superintendent, giving it the force of regulations, and which they implemented the very day that they took up the reins of leadership. In this policy the Class dedicated itself "to develop discipline based upon mutual respect and the principles disclosed to us in the recently established Leadership Course." They further avowed in this instrument their intention "to utilize the methods which eliminate the flagrant violations of mature personal dignity" in attaining and maintaining that discipline. This was a new concept, backed to the fullest extent by the Class which inaugurated it, and by the Administration. It has worked to a degree unanticipated by even the most optimistic. Hazing at the Academy has been prohibited by Congressional act for many years, but, in varying degrees, it has existed throughout those years. This year, to the very best of my knowledge and belief, there has been no hazing. The Class Policy ordained that "private, man to man correction will supplant all practices of hazing," and it has. It also required that "shoving out and other practices of hazing shall cease in the Wardroom Mess," and they have. Plebes are no longer bellowed at by any and all upperclassmen; they don't "swim to Baltimore," "sit on infinity," or suffer any other personal humiliation. The first class have made the indoctrination of the plebes their own responsibility, and a solid bond of mutual respect has grown between the two classes. The second and third classmen are observing, and learning.

Originally, this change put through by the recent first class was designed to protect the plebe from violations of his personal dignity. The benefits from the policy were to accrue to the plebes primarily, with a resultant increase in loyalty and cooperation between and among all classes to be expected. In practice, it seems likely that an even more important benefit has been reaped by the first class in the experience they have gained in treating men as men, in not leaning only on their preponderance of stripes and playing on the fear of punishment to gain results. What lesson could be more important than this? What could be worse than a young officer going out to the Fleet accustomed to gaining obedience by shouting at and, if need be, humiliating his subordinates just to show who is "top dog"? It is common knowledge, reinforced by the experiences of all the Services in the past war, that the American enlisted man can most definitely not be handled successfully in any insulting or condescending fashion. Techniques which embrace a consideration for the man as a man must be employed if mutual respect is to exist and if there is to be any real leadership and any cheerful following. Beneficial practice in such techniques is afforded the first classmen in effectuating their policy and, over the years, the entire Naval Service will be the better for it.

It is hoped that what has been written here will serve to refute the statements of those who would have the public believe that the Naval Academy is still living in the days of sail, doing things as they were done in Dewey's day, and completely out of touch with reality in this changing world. The changes mentioned here have been made, and real progress in the field of human relations has resulted. There is still, of course, room for improvement, and no one is more aware of that than are the officers charged with administering the Academy. All the selection techniques available to the Navy are being utilized, and further perfected in the process, to insure that only those officers most particularly qualified

for career-indoctrinational duties are ordered to these recently emphasized billets in the Executive Department. It is a difficult job, and one
that is administratively burdensome, but noticeable improvements have
already been evolved. Likewise, efforts are being made to work out a program which will afford to the officers selected some formal schooling in
the subjects they will teach as Company Officers. Beginning with the
Academic Year 1948–49 the Company Officers will be teaching to the
first and second class midshipmen not only Leadership, embracing some
areas of Psychology, but also Naval Organization and Administration
and Military Law. It is recognized that some instruction in those subjects,
and in techniques of teaching, should be made available to the officers
selected to do the teaching if they are effectively to discharge their responsibilities. In short, the situation at the Academy today is not static. It is
recognized that only by continuing and consolidating the progressive
changes which have been made, and by holding ourselves ready to make
those changes indicated as necessary in the future, can we say that the
Naval Academy has learned from the lessons of the recent war, and is
profiting from its learning.

Enlisting in the Marine Corps in 1937, **Major Williams** served as
a private for a year on the U.S.S. *Oklahoma* before winning an
appointment to the Naval Academy. Graduating with his class in
December, 1941, in a war-hastened graduation, he was commissioned in the Marine Corps and saw service for two and a half years
in the South and Central Pacific as anti-aircraft battery commander.
Subsequent duty included command of the Marine Detachments
on the U.S.S. *California* and the U.S.S. *Los Angeles*. For the past 18
months he has been on duty at the Naval Academy as Company
Officer and Leadership Instructor in the Executive Department.

"The Fine Line at the Naval Academy"

6

Vice Admiral James Calvert, USN

U.S. Naval Institute *Proceedings*
(October 1970): 63–68

*The contest between professionalism and academic effort fig-
ured heavily in the founding of the school and has been evident
in its affairs ever since. The present Superintendent intends to see
that the balance—"the balance between Athens and Sparta," he
calls it—is retained. But, he reminds us, Annapolis' fame rests on
having produced effective leaders, not renowned scholars.*

ONE HUNDRED AND TWENTY-FIVE YEARS AGO this month about
60 somewhat disheveled young men assembled in an old Army fort on
the Severn River along with some officers and a small group of civilian
instructors. The formation was unimpressive but, nevertheless, important from a historical point of view. The U.S. Navy, after prolonged controversy, had decided to move its officer education program ashore—the
formation marked the beginnings of the Naval Academy at Annapolis.
The conservatism which so often becomes part of the character of men
who deal with the sea was particularly evident in this controversy. How
could young men learn to be naval officers except by going to sea? If the
professors who, in 1845, were a small, but established, part of the Navy,

were permitted to get the midshipmen ashore in some sort of permanent school, who could tell what would happen? Certainly nothing good!

Nevertheless, two developments had come about which made the move appear, on balance, to be necessary for the long-term well-being of the Navy. First, there was a growing need for more professionalism in the Navy's officer corps. In the fall of 1842, a near-mutiny had occurred in the brig *Somers* while she was returning from the African coast to New York. A midshipman (who happened to be the son of the Secretary of War) and two enlisted men were tried, convicted and hanged at sea. The resulting investigation and furor focused unfavorable public attention on the conditions in which future naval officers were being trained.

The second factor was the introduction of the steam-driven warship into the Navy. The launching of the *Fulton*, the Navy's first practical ship of this kind, in 1837, was followed only a short time later by the Ericsson-designed screw-propelled *Princeton*. It became apparent to many that these new warships and their machinery would revolutionize the Navy and place new requirements on its officers. A school ashore was the logical place for the Navy's midshipmen to receive the more complete education needed to meet the challenge.

These two factors so important in the founding of the Academy—military professionalism and the need for sound education—have played significant roles in its history ever since. To some degree they are always in conflict, but they can, and have lived together in more or less fruitful coexistence on the Severn for 125 years. Each factor has its strong advocates, each has its staunch opponents, but again and again it has been demonstrated that the graduates of the Academy who contribute the most are those who have best combined true military professionalism with sound learning. Clearly, this will be the case in the remaining years of the 20th century. There can be little doubt that the Navy faces difficult times in the years ahead. Budgetary considerations and, indeed, the whole mood of the American people indicate that a lean period is in the offing. Historically, the Naval Academy has played an important role for the Navy in such periods. Will this be true again?

In an attempt to find an answer to this question, let us see if we can obtain some sort of perspective on the history of the Naval Academy as it relates to the Navy and to the nation. Taking the risk inherent in all historical generalizations, I would submit that there have been three great cycles in the history of the Academy. Each of these cycles has been characterized by a period of high military and naval activity, a period of change at the Naval Academy, a period of declining support of the Navy itself by the Congress and, finally, a period of high professionalism and productivity at the Academy.

Let me be more specific. The Academy was barely started when the Civil War disrupted it almost completely. When Admiral David Dixon Porter came back from the war to take over as Superintendent in 1865, the old site on the Severn had been almost wrecked through its use as an Army hospital. He and his Commandant, Stephen B. Luce, had to start from scratch in the renovation of the site and the establishment of an effective institution. The late 1860s were great years of change at Annapolis. Porter and Luce renovated the grounds, introduced new uniforms, commenced military drill, competitive athletics, and the system of graduated privileges which has been fundamental to the military side of the school ever since. Discipline, dedication, and accountability appeared on the Severn for the first time and the Academy began to establish a reputation for professionalism.

The 1870s and 1880s saw the Navy itself neglected by the Congress as the nation turned its attention elsewhere. Ships lay at anchor, promotion stagnated, and the profession seemed to be declining—but there was a great period of productivity at Annapolis. The engineering education was improved, military professionalism remained at a high level, and the Academy received international attention for the quality of its programs.

The second great cycle began with the naval activity associated with the Spanish-American War and World War I. Once again, great changes occurred at Annapolis. The new buildings of Ernest Flagg were put up in the early 1900s and the physical appearance of the Academy, as most of us have known it, came into being. Admiral Dewey laid the cornerstone of the new chapel, smart new uniforms were introduced and, in 1907, Anchors Aweigh was played for the first time at an Army-Navy football game. In the 1920s and early 1930s a new wave of anti-militarism struck American society and the Washington Naval Conference was only the first manifestation of another period of neglect and decline for the Navy itself. As many of the present generations know, however, these same years were a time of great productivity and effectiveness for the Naval Academy.

If this historical analysis has validity, it is entirely possible that we are experiencing a third great cycle at the present time. In a broad sense, World War II, the Korean War, and the Southeast Asian conflict add up to one long period of military activity which has had an impact on American society evident to all of us today. There have been changes at the Academy during the 1960s which, while less significant than those which occurred during the 1860s, are nonetheless among the most fundamental ones in the history of the school. As I said earlier, there are signs that we are entering a period of declining support for the active Navy itself. Will Annapolis be able to maintain its tradition of unusual professionalism and productivity during this apparently inevitable period of slackened activity? Only time will tell, but a sense of history is important to any great organization and there are those of us at Annapolis who are confident that the Academy is just now entering into one of its great periods of service to the Navy.

Let us examine, a bit more closely, what has been happening on the Severn during the 1960s. First of all, there have been physical changes. The first major building program since Ernest Flagg's buildings were put up at the turn of the century is underway. San Francisco architect John Warnecke has been given the task of creating the new look at Annapolis and he is off to an excellent start with the recently-completed Michelson and Chauvenet Halls. While strikingly contemporary, they nevertheless pick up the mansard roofs and generally French flavor of Flagg's 70-year old structures in a manner which avoids an unpleasant clash of new and old. More important, the classrooms, lecture halls, and laboratories of Michelson and Chauvenet are among the finest in any undergraduate institution in the nation today.

Warnecke has been given the contract, along with the George M. Ewing firm of Philadelphia, to develop the new library and engineering studies complex which are the next parts of the master plan for the new Academy. The library design is complete, Congress has authorized and appropriated the funds, and the contract for construction was recently signed. The new library is a superb architectural design which, while similar to Chauvenet and Michelson in motif, is sufficiently different to provide variety and symmetry. It should be ready for use in 1972.

The engineering building design is nearing completion and authorization of the first portion of this sizable undertaking is now under study by the Congress. The laboratories of the new engineering complex will be appropriate to the advances in engineering curriculum which have occurred at the Academy during the 1960s and, which indeed, will enable Annapolis to take its place among the nation's foremost undergraduate engineering schools. This magnificent structure, when completed, will be perhaps the most symbolic of all the new buildings at Annapolis. It is built for, and suited to, the advances in engineering education which have been part of the reforms of the 1960s. It recognizes the new requirements for engineering excellence posed by the ships and aircraft of the modern Navy and it symbolizes the fact that, despite all the requirements placed on it for other academic skills, the Naval Academy must always remain, primarily, an engineering school.

In addition to the new buildings, there has been a significant effort in recent years to upgrade the appearance and maintenance of the Yard itself. The Academy has recognized, with full support from its leadership in Washington, that pride in appearance, always important to a military organization, will be particularly essential for Annapolis during the 1970s.

Fine new buildings and handsome grounds will do very little to accomplish our goals, however, unless the young men who graduate from the Academy have themselves the professionalism and academic excellence symbolized by Annapolis in the past and required in more demanding measure by the future. This is where the main effort must go—and has been going. The story of the curriculum reforms of the 1960s has been told so often that it needs no more than a brief summary here:

- In the academic year 1959–60, provisions were made, for the first time, for entering midshipmen to take validating examinations which could exempt them from prescribed courses within the essentially lock-step curriculum and thus free time for elective courses. Electives thus in obtained, along with overloads authorized special cases, could be combined to produce a minor, or even a major, beyond the basic curriculum.
- The Curriculum Review Board of 1959, chaired by Dr. Richard G. Folsom, President of Rensselaer Polytechnic Institute, and consisting of four other academicians plus then Rear Admirals Frederick Ashworth and Horacio Rivero, gave specific recommendations in methods of instruction, laboratory procedures, the library, admissions policy, faculty, and administration which provided, more than any other source, a blueprint tor the academic reforms of the 1960s.
- In the summer of 1963, Dr. A. Bernard Drought, formerly Dean of the School of Engineering at Marquette University, came to the Academy as Academic Dean. Shortly thereafter the position of Secretary of the Academic Board, long held by a Navy captain, was abolished and henceforth all Department Heads (seven Navy captains) reported directly to the Academic Dean.
- In the fall of 1964, the basic course of study for midshipmen, known as the core curriculum, was reduced to become 85%

of the semester hours, with the remaining 15% made available for electives, so that each man could obtain a minor without validating or overloading. Validation and overloading could, of course, be employed to produce further enrichment of the program and, in many cases, a major.

- In the fall of 1966, the then Under Secretary of the Navy, Robert H. B. Baldwin, appointed a 12-member Academic Advisory Board to assist the Superintendent in the conduct of academic reforms. This move was partially in response to criticism of the Academy's programs which had evolved from the 1965 visit of the Evaluation Team of The Middle States Association of Colleges and Secondary Schools. Significantly, two members of the 1959 Folsom Board were appointed—Dr. Folsom himself, still President of Rensselaer, and Admiral Horacio Rivero, then Vice Chief of Naval Operations. Dr. Milton Eisenhower, then President of Johns Hopkins University, was the first chairman, and Vice Admiral William R. Smedberg, a former Superintendent, has served on the Board since its inception.

When I arrived at Annapolis in the summer of 1968 to relieve Rear Admiral Draper Kauffman as Superintendent, he summarized for me his general satisfaction with progress in the academic area, but he also flagged three concerns:

- First, the nearly decade-long effort to upgrade the content and quality of the core curriculum (and, in particular, its analytical and principle-oriented engineering content) had resulted in a serious loss of professional emphasis. No one wanted a return to the trade school approach, but there was such concern over the problem that Admiral Kauffman had appointed a board of five Navy captains to study the matter. He presented me with their three-inch thick report.

- Second, the same long-continued upgrading of the core curriculum engineering content had produced a situation where several of the core courses were so difficult that at least a third of the members of each class were having difficulty in getting more than a plug-and-chug understanding of the material.
- Finally, there was clear evidence that academic performance in the elective courses was far better than in the 85% of the program required as core courses. Both Admiral Kauffman and Dean Drought were anxious to develop some means of increasing the percentage of time spent in elective courses.

With these turnover briefings freshly in mind, I turned to the report of the Professional Education Board. It was clear from that report, from my own recent experiences at sea, and from conversations with many officers of the Fleet, that our main task had to center around improving the quality and quantity of professional, shipboard-oriented education and training at the Academy. The reasons for this conclusion were many but they all added up to the same thing—the mission of the Academy is to produce professional officers for the naval service. In our enthusiasm to be responsive to the post-Sputnik wave of concern over American education, we had permitted professional education and training to be pushed too much to one side.

On the other hand, the Academy was faced, in the summer of 1968, with two rather unpleasant, but highly relevant, facts. The combination of long-term, unprecedented affluence in the nation, coupled with a growing wave of anti-military feeling stemming mainly from the war, had begun to have its effect on the Naval Academy, as well as on West Point and Colorado Springs. The number of nominations from Presidential and Congressional sources to enter the Naval Academy had been on a steady overall downward trend and had reached a point of genuine concern. Secondly, the voluntary resignation rate had been increasing and, during the summer of 1968, was a matter of particular concern with the new Plebe class.

All of these factors had to be taken into consideration in the determination of corrective moves. Obviously, some of the factors were conflicting, and considerable consultation and thought were indicated before any specific steps could be taken. I asked for a thorough briefing by each academic department, including the texts in use, visited classes, talked with many professors and midshipmen, discussed the problem with alumni of the Naval Academy, and talked with selected high school teachers and counselors. The fall meeting of the Academic Advisory Board in 1968 was particularly fruitful. Here I was able to get the views of Dr. Folsom and Dr. Herbert Longenecker, President of Tulane, as well as those of Admiral Rivero and Vice Admiral Smedberg. Later in the fall, I discussed the issues with the Board of Visitors, among whose members at that time was Dr. Edwin Harrison, a 1939 graduate of the Naval Academy and then President of Georgia Tech. I was also grateful that during this period I had the opportunity to hold talks with Admiral Moorer, the Chief of Naval Operations, Admiral Clarey, the Vice Chief, and Vice Admiral Duncan, the Chief of Naval Personnel.

By the spring of 1969, the following conclusions had been reached:

- The professional, shipboard-oriented courses in the curriculum were inadequate in quantity and quality. Their timing was not optimum within the four-year period and, in particular, they did not tie in adequately with the summer programs. Practical, shipboard-oriented, engineering, electrical, and electronic principles were not being adequately or effectively taught.
- The core curriculum, constituting 85% of the required semester hours, attempted too big a task. There was something of everything that "each Naval Academy graduate needed to know," but many of the subjects, inevitably, were covered too superficially to be of real value to the midshipmen.
- The minors programs, while aimed in the right direction, did not provide enough time so that any one area could be studied

in sufficient depth to provide that vigorous academic experience which must be part of a useful undergraduate education.

- The curriculum was not effective in keeping the Naval Academy competitive with other colleges in attractiveness to capable high school men. In today's world, the young men we need most do, indeed, study college catalogs very closely to see what courses and programs are available.

How could we both increase the professional shipboard-oriented programs (we estimated they needed to be nearly doubled) and, at the same time, produce academic programs capable of keeping the Naval Academy attractive to the realistic and better-educated youth of today? The answer was fairly obvious—we had to stop asking "What must every Naval Academy graduate be able to bring to the Fleet?" and start asking "What must every Naval Academy class bring to the Fleet?" The core curriculum had to go and we had to permit each midshipman to select a program which interested him, was oriented to the needs of the Navy, and still left room for the sharply increased amount of professional material being planned.

Once the decision to drop the core curriculum concept had been made, a large amount of flexibility was obtained. The entire program could then be restructured to meet our objectives. Most of the spring of 1969 was devoted to constructing a new academic program around the following guidelines:

- The only courses required by name and number are the professional courses. These comprise roughly one-third of the total program—about double the previous percentage.
- Each midshipman may choose an area of major academic concentration from about 20 available. While the majority of these are in engineering, science, mathematics, and management, there are also majors related to the Navy's interests in international affairs and government.

- Each academic major requires certain minimums in mathematics, science, and humanities. In general these are met by the area elective system—a required number of courses in a given area to be chosen from a list of offerings.

- Each academic major produces enough depth of study in a given discipline to produce a recognized major. Free electives in the major area are available in most programs. The engineering majors are adequate in breadth and depth to warrant Engineer's Council for Professional Development (ECPD) certification. (In order to keep the engineering majors within workable bounds, certain of the professional shipboard engineering courses are omitted, but almost all of this practical material is covered in the engineering major subjects themselves.)

- All engineering, math, and science majors require at least five semesters of mathematics—some more. All of the other majors require at least six semesters of foreign language. Thus, each majors program includes either a rigorous program of mathematics or an in-depth study of one foreign language. There is no easy path.

- Because of the intensity and rigor of the programs, it is necessary for a young man to decide during Plebe summer whether he wants to go into the engineering-math-science area or into the international affairs-government area. Specific majors choices must be made before the beginning of the Youngster (second) year. Changes in majors can usually be handled up through the end of Youngster year.

- Ceilings have been established on all general areas of majors programs in order to ensure that the proportion of graduates in the various disciplines is in general consonance with the Navy's P-coded subspecialty structure, the overall Postgraduate program, and the available studies giving projections of naval officer skill requirements in the years ahead.

The majors curriculum was inaugurated in September of 1969 for the three lower classes. The Class of 1970 had gone too far with the old program to shift over. In September of 1970 all four classes were enrolled in the new curriculum. All of the preliminary indications are favorable. Academic performance is up sharply, the percentage of high grades is up, academic discharges are down significantly, and informal reaction among both faculty and midshipmen is excellent. Moral: the young men of today perform better studying in areas of their choice rather than in a lockstep curriculum.

Performance in the professional courses has, thus far, been encouraging. Reaction to the Youngster LPD dedicated cruise is excellent—for the first time in many years there is genuine excitement among the Plebes in looking forward to Youngster cruise.

Developments in regard to interest in entering the Naval Academy in the summer of 1970 have been noteworthy. In what has to have been the wildest spring in the history of higher education in this country, more young men were nominated for entry to the Naval Academy than ever before in our history. Many efforts were undertaken simultaneously with the curriculum change and it is impossible to say what factors were actually operative. However, in a year during which every other officer-producing program in the Navy suffered a decrease in interest, nominations for the Class of 1974 were up 17% over the year before and 29% over the year before that. We will probably not know for some time what the true reasons were, but it is safe to say that the majors curriculum did no harm.

Important as curricular reform is, it can be only part of the effort to shape the Naval Academy for maximum effectiveness in the closing years of this century. The program for discipline, accountability, and routine at the Academy has been examined carefully to see that it remains tough enough to do the job, but is also up-to-date and realistic enough to make sense to young men coming from an environment significantly different from that which existed ten years ago.

In a time when disciplinary standards are being thrown up to question in all corners of our nation, when respect for constituted authority in civilian society is at an all-time low, and when the national interest in academic excellence remains high, there is great danger that the balance between professionalism and academic activity at Annapolis could be distorted in an effort to "keep up with the times." Many critics of the Academy over the past decade have decried the large amount of time "wasted" on military activities, athletics, the disciplinary program, and the leadership activities of upperclassmen. On the other hand, the counter-critics point out that academics are becoming too important at Annapolis, that the place is being turned into a university, and that dedication to the Navy, military professionalism, the competitive spirit, and discipline are being downgraded.

The contest between professionalism and academic effort is not new. As I said earlier, it figured heavily in the founding of the school, and it has been evident in its affairs ever since. The balance between Athens and Sparta must be retained, but I would point out that Annapolis became world famous as a training institution that produced effective leaders, not as an educational institution that produced renowned scholars. Both are highly important, both can exist together with benefit, but the Naval Academy will succeed or fail in the decades ahead to the degree that it produces professional officers who have the dedication and loyalty to remain with the naval service and do an excellent job.

Another area of genuine importance to the Academy during the decades ahead is athletics. Volumes have been written about athletics at the service academies but only rarely do we get at the heart of the matter. Pride and emotion are an important part of military professionalism. The man whose drives and goals deal only with purely rational matters will rarely find satisfaction in a military career—particularly in a period when civilian esteem of the military is low. Team spirit, the battle cry, camaraderie, heroism, the desperate fight against impossible odds—none of these are very compelling to the strictly rational mind. Competent men who stay with a professional service for long periods of time do so for reasons which are at least partly related to spirit and pride.

The same factors apply at Annapolis. Pride in the Academy on the part of the Brigade is important to the Academy's effectiveness. To the type of young men who, in the main, have been attracted to the Academy over the years, pride in athletic teams is an important part of that feeling. This does not mean that a man need participate himself to share that sense of pride. Quite the opposite, in fact. Who among us will forget how much brighter the world looked the day after that memorable game in 1950 when an apparently hopeless Navy football team stunned proud Army 14–2? And for those who are interested in lacrosse, 8 to 7 may be a famous score for at least a few years to come. It is the great victory against heavy odds that catches the imagination, that inculcates a perhaps unreasonable—but invaluable—pride in the organization.

Unless pride, emotion, and dedication can be generated at the Naval Academy in the disenchanted Seventies, then the Academy cannot fulfill its mission for the Navy during these years. Athletics are only a part of that pride, but they are an important part. The Naval Academy, with teams that consistently lose or play listlessly, cannot be the Naval Academy we need. Restricting our program to intramural athletics only might be more rational, but we need more than rationality for the task ahead. We need pride, enthusiasm, and spirit. Athletics cannot do it all by any means but, like the salt in the stew, they can add a lot to the success of the whole.

These, then, are the major areas of change undertaken during the 1960s—important new buildings and a better physical appearance of the plant, a broader and more flexible curriculum featuring renewed emphasis on professional material, a still-tough but updated disciplinary system, and an athletic program which aims to remain competitive against some pretty big odds.

The balance between the professional program and the academic one is important and has been carefully weighed in an attempt to meet the challenge of the 1970s in both these areas. There is no question that the Navy is becoming more technical, that its weapons, ships, and planes are more complex to operate and to maintain. But it is also true that the graduates of the Academy are going into the toughest leadership environment that junior officers have ever known. They need all the intelligent leadership training they can get. But above all, they must have the pride in the organization, the dedication to its goals, the sense of obligation as specially-trained professionals that will enable them to remain with the Navy in the difficult years ahead.

The preliminary results are good—recruitment is at an all-time high, resignation rates are down, and overall retention rates are up—still, it remains to be seen whether or not the Academy can respond to the challenge of the third great cycle in its history. I sincerely believe that our Academy is the answer to many of the deep problems that will confront the Navy in the years ahead. We must use it wisely if it is to fulfill its potential, but that potential, without doubt, is there.

I am occasionally asked by an important government or business leader, "What is it that is so special about an Annapolis education—why can't we do the same thing with ROTC or OCS for much less money?" My answer is always the same—the word is "accountability." A young man who aspires to accomplish useful work in our society must accept, in one way or another, the penalties and limits of accountability. He must be accountable for the way he looks, for the integrity and sense of what he says, for the appearance of the areas for which he is responsible—whether it is his room in Bancroft Hall, the engine room in a destroyer, or a factory. He must be accountable for the value and effectiveness of his work—he must be willing to stand and take the blame when it is no good.

Whatever its virtues, whatever its faults, the Annapolis system has, through the years, in the great majority of cases, produced an ingrained sense of accountability. Not all of us like it, not all of us think of it consciously, but it is there. It produces a strong tendency to get the job done without spending an undue amount of time pondering the imponderables which surround the task. Does the system produce doers rather than thinkers? I can think of worse indictments.

Doesn't our civilian college and university system produce just as much of a sense of accountability? It is my personal belief that up until very recently, it produced an adequate amount in most cases. But a philosophy is sweeping over American higher education today that produces just the opposite attitude. Authority, tradition, moral values—anything that smacks of absolute values—is looked upon as not intellectually respectable. I believe that the system of balanced education existing at Annapolis today, stressing accountability but offering flexible and challenging education, is going to get increasing national attention in the years ahead.

The post–Civil War period and the post–World War I period were times of great change, professionalism, and productivity at Annapolis. It is too soon to say what the post–World War II–Asian wars period will produce. Perhaps the correct balance among all the factors has not yet been found; perhaps the military and naval professions will have to be more drastically reshaped than we can yet foresee in order to respond to the needs of our society. But to those of us who have been following events closely on the Severn, there is evidence of a new ferment, a new pride, and a new determination. The second 125 years may prove to be more fruitful and exciting than the first.

A graduate of the U.S. Naval Academy with the Class of 1943, **Admiral Calvert** was a submariner during World War II. He commanded the USS *Skate* (SSN-578) when, in 1958, she became the first submarine to surface at the North Pole. He was commander of Cruiser-Destroyer Flotilla Eight prior to becoming the 46th Superintendent of the U.S. Naval Academy in July 1968.

7 "The U.S. Naval Academy: The Legacy of Leadership"

Rear Admiral Thomas C. Lynch, USN

U.S. Naval Institute *Proceedings*
(April 1994): 75–79

LITTLE MORE THAN 30 YEARS AGO, when Admiral Arleigh Burke was the Chief of Naval Operations, he spoke to a roomful of Navy captains who had just been selected for rear admiral. They undoubtedly expected to hear words of high praise for having attained the pinnacle of flag rank; instead, Admiral Burke left them momentarily stunned:

"In the eyes of the Navy," he said, "you are old men!"

He went on to explain that only the most promotable of them would remain in active service for as many as ten more years; most would be gone years sooner. The quarter-century or so of achievement that had brought them to this point served only as prologue to a few brief years when they would have the power and influence to get the things done the Navy most needed doing, before their time on center stage expired. Therefore, their sense of commitment and dedication to the Navy had to be stronger and more intense than ever.

No one could make such a statement with more force and clarity than Admiral Burke. The epitome of honor and integrity, he has always lived for the Navy, and probably could not conceive of any other existence. With unwavering physical and moral courage—in World War II combat and in the Pentagon wars that followed—he set the moral tone for the U.S. Navy, including my generation of midshipmen, then on the verge of graduating from the Naval Academy into one of the most tumultuous periods of American history.

Looking back, I see that my classmates and I probably were more closely attuned philosophically to Admiral Burke's generation of naval leaders than today's midshipmen are to mine—even though my days at the Naval Academy fell roughly at midpoint between Admiral Burke's time and today. Since those early 1960s, the effects of a long and divisive war, along with rapid and profound technological advances and social upheaval and still other wars, have brought about an acceleration of change that makes the generation gap a fact of life rather than a mere sociological concept. It is ironic that such a gulf in understanding between succeeding generations was first foreseen by the masterful science-fiction author Robert Heinlein, himself a Naval Academy graduate. I wonder if he ever thought that his prediction would hit so close to home.

My biggest challenge as Superintendent of the Naval Academy in guiding the total personal growth of each midshipman has been to maintain equilibrium among the competing demands of academic, athletic, professional, and personal development requirements. In the face of continuing change, it is a never-ending balancing act. Change is generally beneficial, but the change we have experienced also has fostered a complacency that takes for granted the principles upon which the Naval Academy—and the nation—were founded.

Clearly, today's midshipmen are bringing to Annapolis far different sets of cultural values and expectations than those of my generation. An instructive case in point has been the perennial issue of falsifying age on identification cards, in order to buy alcohol. This always had been an honor violation, and several midshipmen have been dismissed for it. Those who doctor their cards know that in doing so they are accepting the risk of dismissal; yet I am concerned that more than a few within the Brigade today believe the issue is more of a bureaucratic nuisance than the abandonment of one's personal integrity. The recent investigation spotlight has revealed similar attitudes in a cheating incident, in which too many of the participants managed to rationalize away honor violations— even after admitting misconduct—with little or no remorse. Such a gap between basic sets of understanding about the principle of honor is totally unacceptable, and I recognize that the responsibility for bridging that generation gap does not belong to the midshipmen; it is mine and I accept that responsibility.

Amid all the change, one principle has remained rock-solid: Admiral Burke's concept of committed and dedicated service to the Navy and the nation is intertwined so tightly with personal honor and integrity that they can never be separated. What value is such service otherwise? And what better place to inculcate such values—while providing first-rate intellectual, professional, and personal development—than a service academy? Other high-quality schools may provide comparable or even in some cases superior academic surroundings, but only the service academies can serve as repositories for their services' full array of customs and traditions, ideals and honor—sources of pride for their graduates and of inspiration for their cadets and midshipmen. Many universities might react strongly to evidence of widespread cheating, but how many also would see their way clear to deal with other, more sensitive issues of personal integrity—which the service academies must deal with in no uncertain terms?

Our nation has traditionally and rightly placed the highest expectations on its service-academy graduates, partly because of their demanding training regimen, partly because of their tight focus on issues of honor and integrity, and partly because of their strong sense of commitment. To ensure that Naval Academy graduates will continue to meet these expectations in the future, a Strategic Plan—developed by the officer and civilian faculty, staff members, and midshipmen— was published in June 1992. The Strategic Plan acknowledges that the focal point of all effort expended at the Naval Academy must be to develop midshipmen morally, mentally, and physically, to provide the naval service and the nation leaders of character. Any endeavor that fails to contribute to that end must be evaluated, then corrected or discarded as required. The Strategic Plan is subject to continuous scrutiny, and already has seen several updates and improvements in the three major areas of Academy life:

Academics: For the past quarter-century, the Naval Academy has offered variety in major fields of study, evolving to its current level of 14 technical and 4 non-technical majors. Because of the rigorous core curriculum, which is strong in science and engineering, all midshipmen graduate with accredited Bachelor of Science degrees, as they did before the majors program was introduced. After completing a recent three-year-long review of the core curriculum, we have reduced the academic load somewhat, but midshipmen still carry 17–19 credit hours per semester. There are no easy courses and no easy majors. At the Naval Academy, everyone works but John Paul Jones, resting in his elegant crypt beneath the Academy Chapel.

The Naval Academy's engineering laboratories are the finest to be found anywhere at the undergraduate level. All courses are taught by full-fledged faculty members—not graduate teaching assistants—and extra instruction is always available. The faculty is balanced evenly between civilian professors with doctorates and officers with at least master's degrees in their disciplines—a practice that goes back to the school's founding in 1845. Class size remains small, averaging fewer than 18 midshipmen per section. The Naval Academy has had a Rhodes Scholar for each of the past two years, and other midshipmen have won Marshall, Fitzgerald, Pownall, and Zipf scholarships. Twenty more each year initiate their graduate education while still midshipmen. In the Strategic Plan, the faculty members of the Naval Academy are challenged with the task of taking this wonderful academic system and creating a new partnership between student and teacher, to produce an environment that fosters leadership, creativity, and a lifelong thirst for knowledge in each graduate.

Athletics: Rigorous physical development has long been a staple of Naval Academy life, and will continue focusing on a lifetime of personal fitness. At present, about one midshipman in four participates in varsity intercollegiate competition (18 male and 11 female varsity teams), and every midshipman engages in some form of organized physical activity at the intercollegiate, club, or intramural level each semester. Virtually all intercollegiate varsity athletes and most of the others are involved in year-round strength and conditioning programs.

With regard to NCAA competition, the Naval Academy (and our alumni) must remember that the most important by-products of athletics are the leadership principles absorbed on the field of play—not just pride in having winning seasons and beating Army.

Professional Development: The recent change in this area has been the most significant. From Plebe Summer to summer training to pre-commissioning service indoctrination, the thrust of revisions made has been to emphasize ideal leadership concepts and instill in all midshipmen Vice Admiral James Stockdale's philosophy of "Moralist, Jurist, Teacher, Steward and Philosopher." Because the successful leader must be a selfless person, midshipmen must consider, first and at all times, the impact of each thought, word, or action on their command and those who live and work around them, before considering any personal benefits they may accrue. Leadership training is conducted over a four-year continuum, with formal instruction reinforced by practical experience in running the affairs of the Brigade in Bancroft Hall.

Because most military-specialty training is conducted after graduation, the Naval Academy is able to concentrate on the personal development of midshipmen. At the recommendation of our Board of Visitors, we have established a Character Development Center, headed by a senior Naval Academy graduate—Colonel Mike Hagee, U.S. Marine Corps—who reports directly to me. The Center coordinates all aspects of character development, going well beyond the Honor Concept to include the faculty's continuum of ethics instruction, the Navy's core values program, and our own command-managed equal-opportunity program.

The goal of this unprecedented concentration of effort and resources is to produce high-impact junior officers for the Navy and Marine Corps who will prove inspirational for their seniors, subordinates, peers—and the nation at large. Midshipmen at the Naval Academy today are being held—properly so—to a higher standard than most of them have ever before experienced. Some have faltered; others will falter as they proceed further down the line. But most are thriving as they receive continuous guidance and encouragement to achieve their highest potential.

There are times in life when adversity sounds a wakeup call—and a call for action. We are heeding that call. I have every confidence that Admiral Burke—and the generations of Navy and Marine Corps men and women who share his commitment to integrity, honor, and undying devotion to a life of service—will continue to take pride in the graduates of the Naval Academy and all the service academies. In our relatively short span of remaining active-duty service, my generation will keep working hard to make it happen.

A 1964 graduate of the Naval Academy and its 54th Superintendent, **Admiral Lynch** is a surface warfare officer who also has served as the Navy's Chief of Legislative Affairs.

8 "The Honor Code: Master or Servant?"

Vice Admiral Howard B. Thorsen, USCG (Ret.)

U.S. Naval Institute *Proceedings*
(April 1994): 43–44

THE AMERICAN PUBLIC consistently ranks our military at the very top of the nation's major professions and institutions in terms of trust and confidence, a position earned by more than two centuries of high ethical behavior and strict standards of accountability imposed on those few who transgress.

For the same reason, the four military academies (along with a small number of private institutions with a strong military orientation) have a special place in the public's mind. To be sure, the portrayal of a near-perfect moral and ethical atmosphere is an important, unique differentiator during the highly competitive annual quest by those schools for the very best candidates.

Each year, they accept a total of about 4,000 of this nation's finest young men and women and, in an intense atmosphere, confront them simultaneously with a demanding military indoctrination and academic challenge.

Entrants arrive each summer with their own individual established characters. Highly motivated, they embark on the rigorous routine of midshipmen or cadets. Training and studying from early morning to late at night, they are transformed from civilians to proud members of an elite military organization in an astounding two-month metamorphosis.

They learn how to march and drill . . . and they learn the proud history and traditions of their service. They are introduced to discipline, regulations, and an honor code. Discipline is a broad, dynamic principle of any military unit, and involves far more than the stereotypical "blind obedience to orders" sometimes used by critics intent on discrediting the military. Regulations, on the other hand, guide them in their daily routine within the military organization.

But what about the honor code? That short phrase that states the obvious: that anyone who lies, cheats, steals, or otherwise attempts to deceive is unworthy to remain in the group. What could be more straightforward?

How, then, can it be that violations by large numbers of our nation's finest over the years have occurred not during the first few transition months as they adapt to a new life, but long after they have been accepted into the group and, in some cases, have earned positions of leadership within the corps?

What has gone wrong? Where do we apply the corrections? Of course, we can always find someone to "hang" for letting things slip, conduct a quick study that validates the code, and then go about our business for another decade—confident that the code has been fixed.

Anyone associated with the administration of any of the academies soon learns that the "old grad" solution is invariably more stringent, wishing that less flexible standards of conduct be applied to ever-younger people.

A typical view, perhaps the most common, of what it takes to put things right includes the following:

- An honor code must be strict, strictly enforced, and based on a zero-defect goal.

- Despite diverse backgrounds, young men and women voluntarily brought together by a common goal can and should be treated as homogeneous parts of a whole.
- An intense, high-pressure academic environment offers an excellent opportunity for testing an individual's adherence to the honor code, particularly as it pertains to cheating.
- Those who are living by the code are best qualified to judge the severity of infractions by their contemporaries and determine the punishment, but always with the first principle as the single, unchallengeable precept.

Much of this might have been appropriate 30 or more years ago, but it falls woefully short today, when young officers must be far more skilled as leaders than the junior officers of the pre-1960s. Understanding and appreciating the *concept of honor* is required of all who would lead others, and is directly dependent upon one's own personal experience and stage of maturity. While all candidates achieve parity of authority and responsibility when commissioned, the four years spent at an academy afford an unparalleled opportunity to nurture the moral and ethical standards of the students. To delegate that task primarily to the student body is largely to ignore or waste the chance.

We must consider honor from a broader perspective. Except immediately following those infrequent episodes of cheating, how much effort is expended to develop the nascent apprentices' perception of honor? What tangible evidence to include this vital aspect in the students' education can be found? For example, are all members of the military faculty and staff selected for tours of duty at the academy *primarily* because they personify the high standards we want all graduates to emulate? Do *all* members of the permanent faculty and staff, civilian as well as military, realize that their conduct and approach to duty are on parade every day as part of students' total experience?

The cost associated with the education and training at our academies is considerable. The value of the total experience must prove to be well worth the price. There are no academy courses that could not be duplicated at a civilian institution. There is no civilian institution, however—other than the ones alluded to—that even comes close to duplicating the conditions of discipline and responsibility that exist at the military academies. You have to go to class and you have to shine your shoes, for starters. But you also have to tell the truth because you are being groomed to enter a profession in which lives—not dollars—will depend on your integrity.

At what point is a midshipman expected to have achieved that high sense of honor deemed essential for an ensign? How does that ensign's sense of honor, at that point, compare to those who have a few more crows feet around their eyes and scrambled eggs on their visors?

Is it reasonable to have a strict honor code applicable to college-age youths with totally diverse backgrounds—but then have only a general, unstated expectation for a "nicer sense of duty" from those same individuals once commissioned and serving? Does anyone think that the Uniform Code of Military Justice functions as a tool for raising the standards of honorable conduct, or even heightens the awareness among those subject to it?

An individual's personal standards of ethical and moral behavior—integrity—and acceptance of responsibility for his own action or inaction—accountability—will define his sense of honor. Only the consistent demonstration of a high degree of integrity, coupled with a sincere, forthright acceptance of personal accountability, will meet the high standards of our officer corps.

Admiral Thorsen, a distinguished Coast Guard aviator, graduated from the U.S. Coast Guard Academy in 1955, taught there 1969–1972, and was the Commandant of Cadets 1980 to 1983.

"What Price Honor?"

9

Josiah Bunting III

U.S. Naval Institute *Proceedings*
(April 1994): 44–45

AMERICAN YOUTH ARE TAUGHT, in a thousand ways, to define themselves in competition with those of their generation—in schools, in sports, in the early achievement of public distinction, and the aggrandizement of wealth. Thus students matriculate at competitive colleges having succeeded in a culture that rewards academic success and demonstrated high academic potential. Inevitably, all is reduced to competition and invidious distinction.

In a kind of Gresham's law of schooling, such competition scours ways or stunts the desire to learn for learning's sake, for the "joy of learning singular things"—or the connections among them—and the intellectual self-reliance such things lead to. Students do not pursue personal records; they pursue the highest possible Scholastic Aptitude Test scores (with all the ridiculous and clattering baggage of special courses to prepare to do well on the SATs), the highest possible grades (including absurd markings like 4.1 or 4.2), and distinctions that exalt and reward them above their fellows. Earned admission to famous and competitive colleges is believed to assure admission to the best law and medical and business schools, which in turn leads to . . . etc., etc.

In this context, in such rich soil, the seeds of a willingness to compromise personal integrity in the name of personal gain find a most congenial and nourishing home. Our culture emphasizes competitive academic success, and such success confers the highest prestige. The connections, incidentally, between such success and the distinctions earned in certain competitive sports are multiple, and given recent, and hideous, punctuation in the Tonya Harding–Nancy Kerrigan episode.

Many, perhaps a majority, of American secondary school students cheat. We can see why. But diagnosis does not lead easily to prescription. Our interest is in the various means by which the different institutional cultures in our colleges train or educate young people not to cheat.

Some use fear and shame. Some have single-sanction penalties, in which cheating leads automatically to dismissal, although, given the severity of the punishment, it is likely that such a system may militate *against* reporting the cheater. Others require the observer to confront the thief or the liar and some retain the formal and formulaic archaism, to be written and signed on examination papers: "On my honor I certify that I have neither given nor received aid."

The essential issue is that the roots of character and integrity are planted in childhood, watered, tended, and cultivated in adolescence, and are not reliably capable of effective pruning in early adulthood. Nonetheless, academic programs that implicitly or explicitly emphasize the fecklessness of cheating (in distinction to its wrongness) are, it seems to me, moving in the right direction. Where competition for grades is sharply reduced, cheating must be reduced apace; but the reduction of such competition should stimulate—rather than impede—those who seek a real education.

In an ideal undergraduate setting, which I take to be the small residential liberal arts college—in which professors are motivated fundamentally by the desire to teach and live among the young, in which education proceeds by discussion, in tutorials, in small seminars, and in which understanding is measured in ways that cannot really be counterfeited—

cheating simply is not a useful option. Obviously, there are exceptions; there always will be. But the daily, constant demonstration that the purpose of undergraduate education is understanding, and the beginning of the cultivation of wisdom—not competitive advantage or academic "distinction"—is a potent immunizing substance.

Unfortunately for them, the federal military academies more-or-less reproduce the most flagrantly contributing factors in the national climate of pushing and shoving, lusting and striving, for competitive professional and academic advantage. It is not their intention to do so, but they do it all the same. Then, in exasperation and public shame, they blame themselves for succumbing to the virus that their new recruits have brought in with them and their means of healing themselves are never original or radical.

The time has come to look at the issue in a radical way, by asking the question, "How can we educate midshipmen, cadets, and officer candidates not to cheat, and to live a life of moral and professional integrity after they leave us?" There may be a number of answers; the current code is a laudable ideal. Apparently, for many students, it is just that. It is not an answer.

Mr. Bunting is Head Master of The Lawrenceville School, Lawrenceville, New Jersey. He was an enlisted Marine before he graduated from the Virginia Military Institute as First Captain of the Corps of Cadets and was commissioned in the U.S. Army. A Rhodes Scholar, he served with the Ninth Infantry Division in Vietnam, taught at the U.S. Military Academy and the Naval War College, and was President of Hampden-Sydney College 1977–1987.

10 "The U.S. Naval Academy: Where To in the 21st Century?"

Colonel David A. Smith, USAF (Ret.)

U.S. Naval Institute *Proceedings*
(April 1994): 75–79

IT SEEMS THAT NOTHING has been going right for the Naval Academy during the last few years. Midshipmen are charged with cheating on electrical engineering exams, with sexual harassment, and with disregard for the honor code. Even the football team has been a source of embarrassment.

The basic problem is that the Naval Academy has strayed from its mission of developing young men and women into military officers and leaders, becoming instead an "almost university." The trend toward demilitarization must be ended, and the Naval Academy's programs need to be refocused:

- The curriculum needs to be trimmed and reoriented.
- Midshipmen need to be held to higher standards of behavior.
- The athletic program needs to emphasize martially oriented sports and develop varsity schedules with appropriate opponents.

For those of us who graduated before 1957, the Naval Academy was casually and irreverently referred to as the "boat school" or the "small boat and barge school." The program was simple: a lockstep curriculum that focused on the operation of ships, with some math, science, and general engineering thrown in. The degree awarded was not in engineering; it was a Bachelor of Science.

In 1957, the Soviet Union launched Sputnik, stunning the Free World and the U.S. armed forces. As a nation, we were obviously behind the power curve in fundamental areas of technology—basic math, science and engineering. It was a wake-up call, and the nation responded. So did the service academies. The Naval Academy initiated the first of a long series of changes in modernizing its curriculum. The lock-step program was abandoned. New, more stringent courses were introduced. Previous college credits were accepted, new majors were approved, and a wider range of degrees appeared. Today, degrees are offered in 18 majors: eight in engineering, six in math and science, and four in liberal arts (political science, economics, English, and history).

Has the Naval Academy gone too far in establishing too many degree programs? Today, its academic program differs little from undergraduate programs at most public and private colleges and universities. Consider some sample numbers of degrees awarded by the Academy in 1993: Only 26 (2.4%) midshipmen graduated with majors in electrical engineering, 21 (2.0%) in marine engineering, and 22 (2.1%) in chemistry, while 155 (14.5%) graduated with majors in political science. Majoring in liberal arts were 365 graduates (34.08%), a number significantly higher than the 19.83% just seven years earlier.

These majors figures raise several issues: Should the Naval Academy be competing with colleges and universities by teaching liberal arts to a large portion of the Brigade? How expensive is it to maintain a large number of academic departments, each of which teach a small number of midshipmen? What level of educational quality can we expect from a wide range of academic programs in a relatively small institution?

What can the Naval Academy do?

First, phase out liberal arts degrees. Higher-quality liberal arts degrees are available at lower cost from colleges and universities—public and private. Dropping these programs will allow the Naval Academy to focus on the remaining science and engineering majors, enhancing their excellence. Savings in faculty and staff—thus in dollars—would result. The Naval Academy would no longer be duplicating academic programs available—at better prices—in the private sector.

The focus on a science-and-engineering curriculum is more appropriate for high-tech armed forces that will become even more technical in the future. Graduates of such a program will be better prepared to lead the evolving forces. A solid underpinning in the humanities is essential—liberal arts degrees are not.

Phasing out liberal arts degrees will, of course, reduce the percentage of faculty committed to these programs, as well as department heads and other staff. Faculty members may argue that it will be difficult to hire top-quality academics to teach survey courses rather than full liberal arts degree programs, but in fact there is a buyer's market in the humanities today.

Second, sharpen the focus on technology-based disciplines. Since a technical education enhances problem-solving abilities in a high-tech environment—especially in operational and crisis situations—technology disciplines are vital to the success of the armed forces of the future. Dropping liberal arts degrees and refocusing the curriculum on science and engineering may result in a temporary slump in applicants to the Naval Academy. With an increased reputation for its technology-based program, however, the Naval Academy might eventually receive increased numbers of applications from more technically oriented youth.

Third, reintroduce and emphasize foreign language study. All midshipmen, not just liberal arts majors, should be required to study at least two years of a foreign language. In today's increasingly interdependent world, the Naval Academy needs to move toward, rather than away from, increased understanding of other peoples and cultures. Any complete education should include some knowledge of a foreign language. Such knowledge is both an introduction to different cultures and a basis for later study, as career assignments may dictate. Study of a foreign language improves the understanding of English, as well.

In fairness, it should be pointed out that those days before Sputnik carried their own challenges. Because we all took the same courses, I had the dubious privilege of enduring three years of freshman English—first at a college I attended after high school, then at a prep school, and finally at the Naval Academy. Many other midshipmen also arrived with college credits; some had even completed a four-year degree program elsewhere. Except for their selection of foreign languages, midshipmen in the 1950s followed identical academic paths.

The backgrounds and capabilities of midshipmen were significantly different. Some had graduated from prestigious high schools with rigorous math and science programs; others came from far weaker schools. Many went unchallenged, and some were downright bored. The curriculum was broad, bland, and shallow.

Still, each of us got a basic engineering education with a focus on shipboard systems, from both a marine engineering and an operational standpoint. New ensigns were comfortable when they reported for their first assignments at sea, and this education provided the basis for a continuum of solid ship and operational knowledge for many years of each shipboard naval officer's career. (Marine Corps, Supply Corps, and Air Force officers obviously did not receive comparable benefits.)

What else can the Naval Academy do?

Issues of honor and ethics must be examined anew. Even more important than the overhaul of the academic curriculum is fulfillment of the Naval Academy's mission: "To develop midshipmen morally, mentally, and physically and to imbue them with the highest ideals of duty, honor, and loyalty in order to provide graduates who are dedicated to a career of naval service and have potential for future development in mind and character to assume the highest responsibilities of command, citizenship, and government."

Midshipmen should be held to a higher standard than those in the general culture. Many people today—including elected and appointed officials—seem more willing than in the past to shade the truth or even lie outright. It is no surprise that problems of discipline and ethics continue to be most troubling. The exam-cheating problems—reflecting as they may broader problems in ethics throughout the Naval Academy—are exceptionally serious because of the impact they may have on the character of future officers. The biggest mistake that can be made is to make marginal changes in the ethics program (e.g., in the Honor Concept) and then announce that the problems have been solved.

Discipline, too, becomes less rigorous with each passing year. Recent changes in plebe-year indoctrination have scaled back this basic training to a single summer. A visitor to Annapolis finds midshipmen outside of the Yard at nearly all times of the day and night. Even the Commandant of Midshipmen recognizes the problems and has started tightening some of the behavior standards. Now all hands—not just plebes—are up at reveille; all hands—not just plebes—are in proper uniform.

The Naval Academy should be a military school, not a university with a military program or flavor. Virginia Tech, Texas A&M, the Citadel, and Virginia Military Institute have programs with more military content and higher standards of personal behavior than the Naval Academy.

Finally, make some changes in the athletic and extracurricular program. Even after weakening the schedules, current football and basketball schedules are counterproductive to accomplishing the Naval Academy's mission. Navy can no longer compete successfully with nationally ranked colleges and universities in these varsity sports, especially where athletic scholarships and opportunities for professional sports contracts exist. Winning is important, and winning varsity sports seasons provide many incentives for midshipmen to support their teams and certainly help maintain high morale.

The Naval Academy should encourage and support martial sports, such as rifle, pistol, fencing, water polo, lacrosse, soccer, track, boxing, wrestling, and sailing. Last year, the Naval Academy's fencing and pistol teams lost their varsity status, joining boxing which lost its varsity billing many years ago. One might ask how a military school can discontinue martial sports that are central to the military ethos.

The Drum and Bugle Corps is another example of the civilianization of Naval Academy traditions. Over recent years, the corps looks and performs little differently from a regular college or university marching band, complete with popular music, complicated formations, and waving of colored flags. The corps must be more than a spirited, talented group of young people; it stands for the centuries that armies (and navies) and music have gone together into battle.

A service academy must be a school for the education and training of officers of the armed forces. The course work should emphasize the handling of weapons, drilling, tactics, strategy, ceremony, and leadership of men and women. Further, the instruction must include science and general subjects to accommodate the increasing part played by science and technology in organizing for modern warfare. Midshipmen must learn to behave like the officers and leaders they are striving to become.

It is time for the Naval Academy to become, once again, a truly military institution.

A 1957 graduate of the Naval Academy, **Colonel Smith** served 20 years on active duty. As a consultant, he is working on a congressional study about consolidation of the services' command and staff colleges.

"Critical to Our Future"

11

Admiral Charles R. Larson, USN

U.S. Naval Institute *Proceedings*
(October 1995): 32–37

DURING AN APRIL 1995 address at the U.S. Naval Academy's Foreign Affairs Conference, Secretary of Defense William Perry was asked, "Where will the service academies be in the 21st century?" His quick reply was, "In Annapolis, West Point, and Colorado Springs."

In his 1995 commencement address at the U.S. Air Force Academy, President Bill Clinton stated, "I have seen something of the debate . . . on the question of whether in this time of necessity to cut budgets, we ought to close one of the service academies. And I just want to say that I think that's one of the worst ideas I have ever heard of. It was General [Dwight] Eisenhower as President, along with the Congress, [who] recognized that national defense required national commitment to education. But our commitment, through the service academies, to the education and preparation of the finest military officers in the world must never wane."

All of our senior leaders have voiced strong support for the service academies and their value to the military and the nation. However, if service academies are to remain relevant, the characteristics we desire in the military officers of the 21st century must be defined, and the academies must have a critical role in producing those officers. A foundation of values for our officer corps must be established. These values of honesty, integrity, teamwork, equal opportunity, and respect for human dignity are the very fabric of our society and the principles on which our nation was founded. They are all issues of character, and developing leaders of character is what the service academies are all about.

There has been a recent increase in criticism concerning the value of service academies and relating them to the downsizing of the military, the passing of the Cold War, and the pressure to reduce the federal budget. Questions of relevance and affordability also have been raised. These questions are being addressed in various media, spurred on by a small number of persistent critics.

What all these criticisms and questions have in common is what the critics would like everyone to accept as universal "truths." These so-called "truths" are actually allegations:

- That service academy graduates cost more than other officer accession sources
- That their graduates do not perform better
- That their graduates do not remain in service any longer than officers from other sources

When one looks closely at these so-called "universal truths," something becomes apparent: *They simply are not true.* We need to examine those arguments more closely, discuss the uniqueness of the service academies, and understand why they are critical to the future of our country.

Universal "Truth" Number One: Service academies cost more.

Cost more to whom? Certainly not to the U.S. taxpayer, who ultimately foots the bill for all military training and education. On the other hand, the figures frequently quoted for the cost of a Naval Reserve Officer Training Corps (NROTC) graduate neglect to include the cost of summer training and the amount to administer and run the program. When these are included—as they should be—the costs are nearly equal. Even if one accepts the incorrect premise that the initial undergraduate costs of service academy graduates are higher than that of the Reserve Officer Training Corps (ROTC), these undergraduate costs are minuscule compared to the overall training costs that accompany a military career.

For example, it costs almost $1 million to train a pilot, and a similar amount to develop a nuclear-qualified submariner. If postgraduate training costs and undergraduate expenses are coupled with statistics on length of service, promotions, and selection for command of ships and aircraft squadrons, the return on the investment is much clearer. When costs are amortized over an officer's career, the higher success rate at postgraduate training schools and the higher retention rate for Academy graduates result in the Naval Academy being 50% more cost-effective than any other commissioning source.

Universal "Truth" Number Two:
Service academy graduates do not stay longer.

This perception does not come from any data officials at the service academies are familiar with. Retention rates for Naval Academy graduates exceed all other officer accession sources at every major career decision point. For example, to produce 40 career-designated officers—those with at least ten years of service and selected for lieutenant commander—requires an initial accession of 100 Naval Academy graduates, 140 from NROTC, and 153 from Officer Candidate School (OCS). Furthermore, the last class that was tracked to reach the 20-year point had a retention rate of 41% for the Naval Academy, 24% for NROTC, and 21% for OCS. So that universal truth has absolutely no factual basis.

The mission of the Naval Academy is to graduate people who will make it to the 10- and 20-year points. Then that million-dollar investment in flight training is worthwhile and will pay off because the graduates stay in the service. People from other sources who enter the service, complete their minimum obligation, then leave are not providing the same return on the taxpayer's investment. One should not look at what the service academies are spending any budget year. Rather, the long-term investment value that the academies provide to our country should be studied.

Universal "Truth" Number Three: Service academy graduates either fail to perform better or are indistinguishable in performance from other officer-accession sources.

What data were used to reach that conclusion? The average number of months it took to make lieutenant for graduates of the three primary officer sources. More than 95% of all officers make lieutenant, and all make lieutenant in the same length of time. Therefore, this is not a valid measure of effectiveness of the relative commissioning sources.

Look at other measures that are relevant: e.g., promotion rates to lieutenant commander, commander, captain, and admiral. First, a great majority of officers from other sources leave the service before they are even eligible for promotion to lieutenant commander. Of those officers who do stay in the service, the selection rate for Naval Academy graduates over the last three years for lieutenant commander, commander, and captain has been at least 10% higher than any other source. Early-selection rates for Academy graduates also are much higher for lieutenant commander, commander, and captain. Over the last 30 years, the Naval Academy has produced between 15 and 18% of the unrestricted line officers, and Academy graduates comprise 27% of the Navy captains and 54% of the admirals.

Some critics say that fewer than one-third of all the admirals are service academy graduates because they count doctors, dentists, lawyers, and other such specialists, whom academies do not produce. The Naval Academy's role is to produce officers with actual warfare specialties— that is, surface warfare officers, aviators, submariners, or unrestricted line officers. Of those admirals, 54% are Naval Academy graduates.

The taxpayer should not focus on costs, but on worth—what the service academies in general and the Naval Academy in particular are doing for our nation and our Navy.

The strongest argument in support of the service academies is the "value added" factor, which makes the Naval Academy program uniquely worthwhile.

The Naval Academy's new Character Development Program adds value. It is a four-year, integrated process, in which basic American values and those of the Navy and Marine Corps are strengthened and reinforced. The Character Development Program starts with a 14-hour education program for plebes [freshmen] during their first summer. Eight hours are dedicated to honor and integrity, and six hours address issues of human dignity. These lessons are taught by teams of officers and senior midshipmen, to ensure that the plebes receive the perspective on these critical issues of both the officers and their fellow midshipmen. The climax to this initial development phase is a visit for all plebes to the Holocaust Museum in Washington, D.C., where they can see firsthand the evidence of terrible consequences that can occur when morality and respect for human dignity are lost and loyalty is misplaced.

The program continues into the academic year for all midshipmen, through monthly seminars led by well-trained integrity-development teams consisting of staff, faculty, officers, coaches, and senior midshipmen. These team-conducted seminars are ethics-based, using a formal text and other relevant material. Each seminar has a specific theme, and all midshipmen participate, in groups of about 15. These 90-minute seminars require advance reading and are conducted for all classes in the middle of the academic day—about 1000 on Mondays—not after hours or at some other time that might be interpreted as lessening their importance.

This past year, the program concentrated on components of honor, integrity, honesty, responsibility, and fairness with readings ranging from Plato's *Republic* to Herman Melville's *Billy Budd*. This coming year the topics center on the foundations of honor and choosing the always-more-difficult "what is right." Subsequent years in this four-year effort will focus on living honorably and being citizens who accept the concept of honor as a way of life.

These seminars, although a critical aspect of the overall Character Development Program, will not in themselves fully develop the sought-after core values. These lessons must be complemented and reinforced in all phases of a midshipman's development, with broad-based support coming from all members of the Naval Academy family. To accomplish this objective, an in-place program called "Ethics Across the Curriculum" structures courses in English, history, government, leadership, computer science, and engineering design, to complement the monthly seminars, and demonstrates to the midshipmen that there are ethical components to be found throughout history. Faculty members who have been a part of these integrity development teams have stated that the seminars have sparked more discussions of ethics in all their classes, and that this has inspired them to make ethics and moral reasoning an integral part of their instruction. More important, an environment has been created in which ethical issues have permeated the entire Academy structure, and where ethical issues are introduced daily to midshipmen.

The next phase of the Naval Academy's program will begin this fall, with the introduction of a formal, required ethics course for all third-class midshipmen [sophomores]. Its purpose is to strengthen the midshipmen's backgrounds in the foundations of ethical thought and moral reasoning, so they will be better prepared to participate in both seminars and make moral decisions as they progress through the Academy. This ethics course also will include a portion built around a book of case studies of real-world ethical choices—including situations experienced by junior officers in the fleet—and the results of those choices. Lessons also have been designed that cover respect for human dignity and consideration of others, for use in the existing leadership courses and as part of the Plebe Summer program.

The Naval Academy has just completed a yearlong review of its honor concept, to make it an integral and more meaningful part of midshipman life. It is necessary for people to internalize a sense of honor before they can bring honor to the institution and honor to the profession. The honor concept originally said that midshipmen will not lie, cheat, or steal. Now, it still says midshipmen will not lie, cheat, or steal—and they also will do what is right and honorable. The education program is focusing on doing what is right and honorable, after defining "right and honorable" and showing how such choices are made.

Complementing these education efforts is a newly established peer support program known as the Human Dignity Education Resource Officer (HERO). Midshipmen in this program are elected by their peers, then formally trained on issues such as conflict resolution, ethnic and gender discrimination identification, harassment prevention, and alcohol abuse indicators. Their purpose is to provide an alternative reporting or identification channel for those sensitive issues that midshipmen may be reluctant to bring to the formal chain of command—and to ensure that the affected midshipmen receive the proper support or counseling they need to achieve their full potential. In addition, the experience gained by these HEROs will be very beneficial when they face similar problems as officers.

This constitutes the Naval Academy's Character Development Program. Where else in the world can a four-year character development program be implemented as effectively? Challenge a hundred university presidents to put together a package like this and teach it at this level of thoroughness and intensity. But we know this can be done at the Naval Academy-and we are doing it.

The second thing that is unique to the Naval Academy is an intensive four-year leadership laboratory-learning how to subordinate self to team, self to group, self to mission, and self to nation; learning how to follow before leading; learning small unit leadership; moving up to larger group leadership; and then preparing for leadership in the fleet.

The entire leadership curriculum at the Academy has been restructured. The changes will allow leadership training to bring the focus back to the basics—the fundamentals that have brought success to the Academy over the past 150 years. Having strayed a bit too much from the basics, midshipmen now need to draw more from real-world case studies. Today, students hear from people who have been out where the rubber meets the road. In addition to learning some of the fundamentals of leadership, the midshipmen return to some of the basics—such as etiquette, manners, and naval traditions. They learn how to treat troops; how to command; how to deal with people; and what makes successful people and leaders.

Material taught in the classroom can be applied to situations found everywhere else at the Academy—in the dormitory, on the athletic field, and in the extracurricular-activity programs.

The Naval Academy produces 15–18% of the unrestricted line officers who enter the fleet each year. These officers have high standards of character, integrity, leadership, and professionalism, along with a sense of the traditions of the Navy and the nation. By example, they can set a standard that like-minded officers from other sources will want to rally around. At the same time, however, the NROTC and OCS programs produce many truly fine officers—so it would be totally inappropriate and wrong to say that service academies are better, elite, or superior.

What the service academies can instill in their graduates is reflected in the remarks of a now retired Marine Corps major general, who earned his commission through the Platoon Leader Candidate program. I had become friends with him during flight training in 1958. During a recent social event, the general said, "Thank you for going back and taking over the Naval Academy. I hope you will produce the kind of people you did back in 1958." Asked to explain his comments, the general elaborated. "I always looked up to you Naval Academy graduates. You showed me leadership; you taught me the way; you set the standard; you showed me the example; you got me started; and I really am grateful for your getting me headed in the right direction."

It is very important to understand that all of the sources will produce outstanding officers, and they all have the potential of producing the Chairman of the Joint Chiefs of Staff, service chiefs, and flag and general officers. However, if the academies are doing the job right, then they will send officers into the fleet who will lead by example and display the standards of integrity, leadership, and professionalism that will encourage like-minded officers to gravitate toward and emulate them. These young leaders need to be dedicated, competent, and prepared to lead by example. There is clearly a need for *excellence without arrogance.*

Service academy graduates can have an enormous impact on the professionalism and the integrity of not only the military services but also the communities in which they live. If successful, Naval Academy graduates and other officers will form one team and move forward together. By the time they are lieutenants, no one should be able to distinguish—by performance—an officer's commissioning source, because they will all be doing well. If this can be influenced by the standards set at the Naval Academy, then that is the success story. That is what makes the value added, makes any perceived extra cost worthwhile, and puts the debate on the proper plane.

Throughout history, support for the military has shown ebbs and flows, with ROTC units being shut down during the Vietnam War and a number of ROTC units being threatened with shutdown over other political issues. The administrations of these latter schools, for example, have said in effect, "If you're not going to change certain policies in the military, then we are not going to have ROTC on our campuses." In this time of uncertainty with respect to the future and in this time of downsizing, it is more critical than ever to maintain this core—this cadre, this input of professional officers—to help us ride through whatever turbulence lies ahead. In a sense, the academies are almost, in a sense, a counterculture, in that they are going against some of the norms of society with this emphasis on character development. Many young people today have an attitude of "Well, as long as you get away with it, it is okay;" or an attitude that says, "As long you don't break a law or go to jail, you're okay." There is almost an aura of twisted respect for those who manipulate the system and work around the edges to achieve success.

This has been spelled out in a letter from a frustrated professor at a leading graduate business school. After attending a faculty and student forum at his university on responsible business practices, the students who participated in the session were demanding that the school introduce problems with ethical content into the Master of Business Administration curriculum. Ironically—maybe even sadly—six or seven of the faculty panelists said that it was inappropriate. Although the students claimed they would be unprepared to contend with ethical problems in the workplace without some preparation from the business school, the panelists said this aspect of the students' lives was not the faculty's responsibility. Here are some of the faculty members' remarks to the students:

- "If you want ethics, go to Sunday school."
- "Don't mix your two facets of life, business and ethics."
- "You cannot teach people to be good people—just good managers."
- "Just obey the law. If there is no law against it, you may do it."

In a letter to his nephew in 1787, Thomas Jefferson said that the moral sense, or conscience, "is as much a part of man as his leg or arm." The Jeffersonian view is the one the Naval Academy and the Navy subscribe to. The view also forms the foundation that makes the Academy such a unique institution. If there is any single theme stressed for the midshipmen, it is the one of integrity. They are taught repeatedly that the most absolutely critical thing in the Navy is integrity. The one thing they can control—the one thing no one can take away from them and the one thing no one else can influence—is their integrity. Midshipmen are told that they must work to keep it, because if they lose it, they are not wanted in the Naval Service. The goal is to make any issue of integrity instinctive and comfortable to deal with—to make any decision to do what is right second nature, even though it often may be more difficult.

The Naval Academy may indeed run counter to the prevailing culture in some ways, as it continues to produce people who are morally, mentally, and physically sound; in doing so, it performs a service for the country. The four-year immersion process at the Naval Academy lends itself to superb personal development.

The continued health of the service academies is as critical today as any time in our history. The cost-effectiveness and the value-added factors are already there. In this time of geostrategic uncertainty over many politically unsettled areas of the world, coupled with today's declining social values, the academy input is needed more than ever. The nation needs a core of strong, young leaders of character who are dedicated and will remain loyal to the principles on which this nation was founded.

Each commissioning source has a significant role in the development of our officer corps; each is very important to our nation's defense. All of them are needed. To be successful, everyone must work together, and good leadership can indeed bring everyone together. Finally, the service academies provide a source of stability that is very important not only to the military, but to society as a whole as we move toward the 21st century.

At the Naval Academy, senior leadership is committed to doing everything it can to produce young people who will lead by example, set the standard of professionalism and integrity, demonstrate excellence without arrogance, and be the leaders of today and tomorrow. American values are alive and well at the nation's service academies, and their leaders intend to keep it that way!

Admiral Larson is currently in his second tour as Superintendent of the U.S. Naval Academy. Both a naval aviator and a nuclear submariner, he was the 15th naval officer to hold the position of Commander-in-Chief, U.S. Pacific Command, before his return to the helm at Annapolis.

12 "The Academy Could Learn a Thing or Two from the Ivies"

Steve Cohen

U.S. Naval Institute *Proceedings*
(July 1999): 50–57

"NOW HEAR THIS: Your new wingmen are Ted Turner and Jane Fonda." Relieved that this is apocryphal? That's not surprising. The next pronouncement, however, is not fantasy, and whether or not it is disturbing lies at the heart of the matter: The U.S. Naval Academy has more in common with the "liberal, elitist, anti-military" Ivy League than it has with any other institutions.

This connection supersedes any affinity the Naval Academy has with the Church, the Republican Party, and maybe even varsity athletics, which more than 80% of midshipmen have participated in before descending on Annapolis.

Certainly, stark differences distinguish these institutions. But their similarities—the young people they attract and the leadership roles expected of them—outweigh the differences. Those parallels provide each with an opportunity to learn from the other and thereby better fulfill its own mission.

In the interest of full disclosure: I was booted out of the Naval Academy halfway through my second-class year. The ostensible reason was misconduct; I was accused of cutting class. But the real reasons are more murky, lost in the fog of selective memory and embellished by too many drinking stories. Just say I was neither sinner nor saint. I always have valued my Academy experience and have enormous respect for the Navy.

Recently, I began to harbor a lingering suspicion that the Academy was not quite working. Too many news reports had surfaced too many serious incidents—cheating, drug, and sex scandals—to make me wonder whether some bad circuits might be corrupting the motherboard.

After three separate visits to Annapolis, several dozen interviews with midshipmen, faculty, and administrators, I have to admit I was wrong. The Academy, far from broken, is working quite well—at least in large measure. It is producing dedicated, capable young men and women who are committed to or are at least opened-minded about a career in the sea services. But could it do a better job? My answer is a resounding "yes."

I do not think the Academy is fulfilling its mission, which is, in part, "to provide graduates who . . . have potential for future development in mind and character to assume the highest responsibilities of command, citizenship, and government." Evidence suggests that the Academy has assigned—intentionally or not—this part of its mission a distinct back seat. Remember the bumper sticker sold at the Midshipmen's Store and the Visitor Center? It read: "Annapolis: America's leaders of Tomorrow." For generations, the Academy had been an incubator of leaders. Today, it seems to have turned inward, virtually abdicating its role in providing the United States with its leaders in government, business, and civic institutions.

What Works—In Part

As a source of career and senior Navy and Marine Corps officers, the Academy is more than holding its own. Captain Glenn Gottshalk, U.S. Navy (Retired), the Academy's Director of Institutional Research, points out that, to produce 100 Navy captains, the Academy has to graduate 790 ensigns. By comparison, Naval Reserve Officer Training Corps

(NROTC) programs have to produce 1,226 ensigns to generate 100 captains, and the Officer Candidate School (OCS) program has to commission 1,587 ensigns to yield 100 captains. In short, through a combination of aptitude, dedication, performance, and perhaps, as some would argue, a touch of favoritism, Academy graduates are more than twice as successful at reaching the coveted O-6 rank as their civilian-school counterparts.

Critics have argued that, while the Academy does generate more career officers, it is not as cost-efficient as other sources. It costs some $200,000 to educate a midshipman at the Academy, compared to only $50,000 in scholarship fees per NROTC graduate. The Academy argues that the disparity is offset once the cost of pilot, nuclear-power, or other specialized training is included in the calculation—a conclusion with which I agree. In business terms, the "total cost" of the Academy-trained officer is about 14% more than NROTC officers. But with higher retention, the "lifetime value" of the Academy officer far offsets that higher initial investment.

Three troubling questions emerge from these seemingly encouraging retention statistics:

Problem One—Attrition: Why is the Academy producing only about 790 ensigns annually (plus approximately 130 second lieutenants for the Marine Corps)? For the last 20 years fully 24% of every Naval Academy class has failed to graduate. This compares with an attrition rate of less than 2% for the Ivies.

Approximately 5% of each Academy class flunks out; just more than 3% is separated for misconduct or honor violations; the balance—about 15% of every class—decides the Academy just is not for them. One senior officer objected to my use of the phrase "failed to graduate." He argued that the decision to leave the Academy and pursue an education and career at another institution is not a "failure." Rather, it is an individual decision reached consciously and with great forethought.

This officer is right: it is not a failure on the part of the midshipman who leaves; it is an institutional failure. Any institution that has so many qualified applicants competing so vigorously for an appointment, and then selects so carefully from among them, should not be losing a quarter of its students.

If, as one officer suggested, this attrition is "planned"—a weeding out of people not right for naval service—it is a fiscally irresponsible policy. Even though most of the separations occur during a midshipman's first two years at the Academy, this attrition is costing the Academy—and the taxpayers—approximately $20 million annually.

But clearly, this is not just a financial concern. Rather, it is a problem that touches on bigger issues and suggests a serious disconnect between the Academy and society. Three possible explanations for this attrition are:

The Academy's own admissions process is misjudging a candidate's potential to do the work or live within the high, clear standards of character and behavior. That admission process is the Academy's own. It involves congressional nominations but is no longer dominated by it. Instead, the admissions formula centers on a "whole-person multiple" that, in theory, balances a candidate's ability to succeed academically at the Academy and, later, as a naval officer.

Young people are attending the Academy with a misperception of what it is really like. Or they are attending for the wrong reasons, the most common of which is parental pressure/expectations.

The Academy is turning off good kids—kids identified as having both the skills and aptitude for naval service but who become disillusioned by the Academy itself.

Problem Two—The Missing Voice: The fleet has eroded to 323 ships. Problems with retention are debated in the news magazines and on the op-ed pages—not just in the professional military journals. Roper reports that the Navy is seen by the public as the least "effective at fighting and winning the nation's battles," and that almost as many people believe that military officers are overpaid as underpaid. Who must be accountable, if not the officers who lead the Navy?

Unsurprisingly, a high concentration of Academy graduates are among the Navy's top ranks: 41% of current flag officers. Certainly, although an admiral's perspective is fashioned by 25-plus years of active duty, the Naval Academy experience provides the foundation. The problem appears to be a missing—or overwhelmed—voice of military leadership in the Navy's decision-making equation.

Many of the officers I interviewed place the responsibility for the fleet's erosion and retention problems with civilian decision makers, not senior military officers. Officers, they argue, advise and recommend but do not make policy decisions. They are the product of a culture that responds to authority—including civilian authority making "bad" decisions—with, "Yes, sir. Can do, sir."

I do not quibble either with the constitutional underpinning or the political reality of this position. I do think, however, that the military's voice has been muted in these policy battles. When the pro-military *Wall Street Journal* criticizes the Joint Chiefs of Staff for their lack of candor about preparedness—as it did last September—one has to worry about how many other policy decisions were made because military officers could not find their voice.

Problem Three—Lost at Sea: In leadership positions outside the Navy, Academy graduates are scarce. Of 535 members of Congress, only one Naval Academy graduate is among them: Senator John McCain (R-AZ). Of the Forbes 1,000 Top Corporations, only five are headed by Naval Academy graduates. Where are the Academy graduates who today are assuming the "highest responsibilities of command, citizenship, and government?"

So why should we be concerned about the Naval Academy? It may be the occasional (one even could say aberrant) scandal—cheating, drugs, sex—that triggers disproportionate media attention. But it is the systemic deterioration of the fleet—both personnel and ships—that demands continuous, objective reassessment of the Academy's practices. It is time for the Academy itself to ask: how can it fulfill its entire mission more successfully? And what can it learn from the Ivy League, which produces so many U.S. civilian leaders?

The Academy's Strongest Asset

On paper, midshipmen are clearly impressive. Their grades, Scholastic Aptitude Test (SAT) scores, and extracurricular activities are almost indistinguishable from their Ivy League counterparts. The average math SAT score for midshipmen at the 25th percentile of the entering class is 630, almost identical to Cornell's 640 and just a bit lower than Brown's 650. Similarly, on the verbal SAT, the midshipmen more than hold their own: the Academy's average score of 600 is ahead of Columbia's 590, and just behind Cornell's 610.

According to the *Time Magazine/Princeton Review* guide to colleges, high school class standing is a bit higher for students entering the Ivies than for those starting the Naval Academy. But in terms of leadership positions held, participation in extracurricular activities, and varsity sports, the difference is negligible. Geographic representation and racial diversity are almost identical between the two groups, and only the percentage of women in each class is significantly different.

Are there other, more-difficult-to-detect cultural or political differences? Are midshipmen more conservative politically? Are the academies more Catholic and the Ivies more Jewish? Is it safe to assume that the incidence of military service among parents of midshipmen is higher than among those of Ivy League students?

One officer I interviewed hypothesized that most midshipmen come from homes where "fathers and mothers live in the same house together, and where the mom is a secretary and the dad is a plumber." I do not know whether the incidence of single-parent or professional families is higher in either population. But I suspect that officer will be surprised to learn that, at Harvard, 78% of students receive need-based financial aid, and 75% of undergraduates hold work-study jobs on campus. (In fairness, I should note that the other Ivies, with smaller endowments, can offer financial aid only to about 50% of their students.)

Some differences are obvious between the students at the Ivies and at the Academy, not the least of which is the midshipmen's willingness to accept the harsh restrictions placed on daily life. When you meet midshipmen in person, their paper-based statistics and accomplishments seem glaring understatements of their decency, energy, and commitment. Immediately clear is the fact that these are among the very best and brightest kids in the country today. I found it interesting that many whom I interviewed said they considered and were accepted by Ivy League schools.

I came away from my midshipmen encounters with a very clear conclusion: these are the Naval Academy's strongest assets. Overwhelmingly, they are honorable, smart, extraordinarily hard-working, and eager to serve—indeed, willing to give their lives for—their country.

Unfortunately, based on my interviews and from the institution's own survey data, too many midshipmen are disillusioned by the gap between the Academy's rhetoric and its practices; practices that fritter away the good will, aspirations, and trust of these terrific young people. To remedy that, and better to fulfill its mission, here are nine lessons the Academy can learn from the Ivy League:

Lesson One: Trust Them

Even though the Academy attracts some of the very best young people in the United States—kids who embrace the honor concept overwhelmingly—the Academy administration demonstrates, in big things and small, that it does not really trust midshipmen. This is manifest in the choices midshipmen are allowed—and not allowed—to make.

The sad irony here is that midshipmen in turn unfailingly recognize this lack of trust, such that an extraordinary cynicism permeates the brigade. In a conversation with me, even the Commandant of Midshipmen,

Rear Admiral Gary Roughead, recognized and admitted to it. This cynicism leads to two types of widespread behavior: First, midshipmen spend an inordinate amount of time trying to beat the system; second, they subscribe overwhelmingly to the principal (unofficial) rule of Bancroft Hall: "Never bilge a classmate."

The most vexing problem for midshipmen stems from the conflict between these attitudes and behaviors and their underlying belief in the honor concept. Every internal survey underscores midshipmen's belief that honor "sets them apart." In more than a dozen off-the-record interviews with midshipmen, they revealed a keen awareness of the ethical and behavioral distinctions between the honor concept and the conduct system. The former is sacrosanct; the latter is fair game. This coincides with the Government Accounting Office's finding of midshipmen's widespread lack of confidence in the equity of enforcement.

In 1996, in direct response to the Electrical Engineering cheating scandal—and, as one officer suggested, in the shadow of "officers throughout the fleet being hammered for (questionable) ethics violations"—the Academy recognized and began to address the underlying problems. The Superintendent, Admiral Charles Larson, improved the character development program significantly. Today, all midshipmen participate in monthly "Integrity Development Seminars," take a core ethics course—"Moral Reasoning for Naval Leaders"—and hear from a wide range of noted speakers on ethical issues.

The changes in the character development program were a very good beginning, but the Academy needs to go further. A similar initiative is necessary to rethink and reshape the conduct system. Here, perhaps surprisingly, the Academy can learn a lesson from the Ivy League.

How do the Ivies demonstrate trust? "They treated us like adults" was the phrase heard repeatedly from Ivy graduates. In part, that translated into a two-pronged approach of "less is more" and "benign neglect."

The Ivies institute few formal rules of behavior beyond the civilian criminal statutes, and they turn a blind eye to most youthful indulgences.

The real demonstration of trust, however, comes from the issues over which Ivy League students must take responsibility. Academically, Ivy League students must choose a much wider range of their courses than their Academy counterparts. (Within the Ivy League itself, this degree of autonomy takes many different shapes. At Brown, no course or distribution requirements exist; at Columbia, a core curriculum is tightly defined.) Ivy students must decide how to structure a major, when to study—even how to study—and whether to keep up with lectures and readings or cram during reading week.

As for social and extracurricular activities, they must decide what to play and when to play, whether or not to engage in sexual activity, and how to handle the freedoms afforded them. The underlying message, however, is that students are adults, they will be treated as such, and they will be held accountable for their performance. But few rules intervene.

Can the Academy adapt—if not adopt—the Ivy approach? Is it unrealistic for the Academy to scrap a reg book that defines midshipman life from reveille to taps and replace it with a set of clear objectives about what really matters? Is four years of seven-day-a-week, 24-hour-a-day micro-management really necessary? Or could the Academy specify what must be achieved to graduate and be commissioned; and then allow midshipmen more flexibility and responsibility to meet those goals?

More important, can the Academy ask these questions and engage the midshipmen in answering them? That alone would go a long way toward establishing an environment of mutual trust.

Lesson Two: Build Self-Discipline

Midshipmen are busier and engage in a wider variety of activities on any given day, month, or year than most of their Ivy League counterparts. The problem is that they accomplish this through extraordinary structure. What happens when that structure is removed, as it is in much of the rest of the "real world?" Midshipmen need to learn to operate successfully without imposed constraints.

When presented with this hypothesis, Commandant Rear Admiral Gary Roughead, my former classmate, argued that midshipmen do have structure removed and often "bump into the walls learning where those constraints are."

Indeed, he is right—but only in part. Midshipmen do have to exercise enormous self-discipline in order to navigate the shoals of daily life and the conduct system. The issue is one of scale and consequence. Whether to fold one's underwear in the prescribed manner—or a hundred other prescribed behaviors—or get "fried," should not be the test of self-discipline for our next generation of leaders. The Academy should be asking: What experiences best test a midshipman's ability to handle lack of structure? And how important are these experiences?

Perhaps the most telling incident of this conflict between self- and imposed discipline occurred during my interview with two of the Academy's academic deans. I was looking for an example where the stakes were higher than whether or not to shine one's shoes. I asked, "Why not allow midshipmen to decide whether they should attend or cut a particular class? Allow them to prioritize, to make the decision that some other demand takes priority over attending class."

The deans obviously were distressed by the question. Their first answer was transparently disingenuous: "Well, the taxpayers are paying for them to attend class." The disbelieving look on my face triggered a more candid response: "Too many midshipmen would probably take advantage and flounder," they feared.

That is precisely why midshipmen must be given the opportunity to learn the skills of self-discipline before they leave the constraints of the yard. Once they become junior officers, the stakes will be much higher.

The specific reforms the Academy can adapt from the Ivy League go beyond allowing midshipmen to decide whether to attend or cut class. Academy-appropriate reforms will emerge if the Academy amends its attitude and approach to change. It should be asking itself a fundamental set of questions, and, just as important, engaging midshipmen fully in the effort to answer them. What is mission-critical, and what is a holdover from another era? What is essential to the character and performance of successful naval officers? What traditions deserve to be maintained, and what fears of becoming "too much like a civilian school" can be ignored? Does the Academy really need daily multiple formations and inspections? Does limiting evening and weekend access to town, cars, or televisions really engender a stronger commitment to studies?

A shift from dependence on imposed discipline to self-discipline can be achieved if the Academy recognizes that self-discipline is a corollary of trust and must support the twin pillars of responsibility and performance. If people believe they are trusted, given the responsibility to earn that trust, and held accountable for it, then a higher standard of performance can be sustained. If people believe they are not trusted, they respond accordingly: They perform for the monitor, typically cutting corners when they believe no one is watching.

Lesson Three: Close the Gap with Civilian Society

Much has been written about the increasing distance, the "disconnect," between the civilian and military segments of society. Although virtually none of the anti-military sentiment that ripped the country apart during and after the Vietnam War still remains, the end of the draft has limited the contact between segments of our society and the military. Probably the most extreme example of this disconnect centers on the Ivy League, which produces so many of our country's business, government, and opinion leaders. (To be sure, this is more the fault of the Ivy League, which banned ROTC from its campuses.)

Unfortunately, this isolation has been exacerbated by changes in Academy life in the aftermath of the cheating, sex, and drug scandals. In 1995–96 then-Superintendent Admiral Larson instituted new restrictions with the well-intentioned goal of getting midshipmen to "focus inward." Rather than allowing upper-classmen to escape the yard at every available opportunity, the admiral wanted midshipmen to spend more social and liberty time engaged in activities at the Academy, thereby building Brigade esprit de corps and cohesion. It was a reasonable objective, except for one long-term unintended consequence: More restrictive regulations result in less midshipman contact with their civilian counterparts, less understanding of civilian norms, and a greater disconnect.

To reduce this disconnect, the Academy can adapt four practices from the Ivy League:

- Make the Academy the place people want to be. If the Academy does not want to precipitate mass-exodus on weekends, it has a choice: It can continue to restrict weekend liberty. Or it can make the yard a place where midshipmen want to socialize with fellow midshipmen and invite civilian friends to join them. Unfortunately, for most midshipmen, the Academy's unrelenting rules and restrictions outweigh its beauty and resources. The Academy needs to find a better balance between its military underpinning and its desire to keep midshipmen from wanting escape every chance they get. Bancroft Hall need not turn into "Animal House;" but neither should it feel like the "Big House."
- Establish an extensive two-way exchange program between the Academy and top civilian schools. Encourage many midshipmen to take a semester at a civilian college, as early as second semester of third-class year. Let them experience the "fun" of a civilian college—along with the more mundane aspects—and require them to exercise newly honed self-disciplinary skills. In addition to a full academic load at their host school, participating midshipmen should be required to fulfill professional-development and "ambassadorial" requirements. Odds are, for the vast majority of midshipmen who take part in civilian exchange programs, the grass of the "outside world" will not be greener.

The argument against such a program will center on the claim that midshipmen have so many additional military requirements to be fulfilled at the Academy. That is a specious argument that can be addressed with a bit of flexibility and creative scheduling. And the benefits are well worth the effort.

- The corollary of allowing midshipmen to attend top civilian colleges for a semester is encouraging Ivy League students to attend the Academy for a semester. Odds are that more than a few— and certainly many more than are now coming from Ivy League schools—would choose to enroll in OCS programs upon graduation. Moreover, the exposure to just a semester of military people, thinking, discipline, and honor would be the single greatest bridge between the civilian and military segments of our society.

- Expand the professional development of midshipmen to include internships with members of Congress. Civilian control of the military is understood intellectually by most midshipmen. It is mitigated, however, by a barely concealed contempt too often expressed by mid-level officers over the fact that the vast majority of both the Congress and the Executive branch never served in the military.

Lesson Four: Raise the Academic Bar

Much has changed academically—almost all changes for the better— since I attended the Academy. The faculty has improved, with virtually all civilian professors holding advanced degrees and being encouraged to pursue research in addition to their teaching duties. The curriculum is broader and richer, with midshipmen taking more humanities and social science courses than they did when I was there. And many more graduate school opportunities are available both to Academy graduates and firstclassmen. Indeed, about 9% of the Class of 1998 pursued advanced degrees.

But after sitting through classes, talking to faculty and midshipmen, and examining specific course curricula, several things are clear:

- While the academic program is rigorous, the process of education at the Academy makes it much more like the nation's toughest high school than an Ivy League college. With compulsory attendance at every class, weekly quizzes and tests, and reading lists and assignments that are far from overwhelming, midshipmen are spoon-fed bite-sized pieces of knowledge that increase the likelihood of their success. It may be rigorous—more like medical school than college, noted one Harvard and Harvard Medical School graduate—but the academies just are not in the same intellectual league as the Ivies.

- Too many midshipmen still spend too much time sleeping in class. In one freshman seminar I sat through—with a professor whose command and engaging presentation of the material was every bit as compelling as most of the Ivy League professors I have encountered—fully one-third of the class drifted off at one point or another. Such behavior cheapens the academic environment. If sleep deprivation is a physical reality for plebes, we should be concerned about what else plebes are missing during that year. If it is not, then we should be furious about the message classroom dozing conveys.

- More than a few officers raised the question, "What difference is there between a midshipman sleeping in class and a midshipman cutting class to sleep in his room?" The answer, I suggest, is respect—respect for the professor and respect for the academic process.

- The Academy experience is one of extraordinary breadth, often at the expense of depth. Typically, midshipmen take six courses per semester, plus physical education, leadership, honor, and character-development seminars. The typical Ivy League student carries only four courses per semester.

- The Academy is not anti-intellectual; but neither is it a hotbed of intellectual curiosity. Compare virtually any equivalent courses, and what immediately becomes clear is that the reading lists and writing requirements for the Ivy courses are noticeably tougher. Few Ivy students read all the required and recommended lists.

But the bar is set higher, and the expectation is that they will come closer than their Academy counterparts.

These concerns about the Academy's academic standards and approach are countered by two very simple—and at first seemingly valid— arguments. First, officers argue that students at Ivy League schools have only one bar to jump over—an academic bar—while midshipmen have so many more. Even civilian students who play varsity sports, volunteer in community service projects, and hold down work-study jobs still do not have as much required of them as midshipmen. Second, the officers argue that life at the Academy is a zero-sum game: if we put more emphasis on academics, then something has to give.

Initially, these arguments appeared hard to rebut. After reviewing the notes of my interviews and re-reading the reports and speeches I had gathered, I concluded that both the administration and faculty want the Naval Academy to be a first-rate educational institution. Evidence of a commitment to excellence was clear. Never did I hear mitigating phrases designed to provide an escape hatch. As one professor put it, "We are committed to giving midshipmen a first-rate academic experience; not a first-rate experience within the constraints of the mission."

Underlying this commitment, however, was a clear recognition— almost a resignation—that the system and environment conspired against the higher standard. As one senior academic put it, "There is a very pragmatic quality to education and midshipman life. Midshipmen often demand of the faculty, 'What do we have to know?' In turn, we have to ensure that we don't just give them the 'gouge' but rather sell them on the fact that the ability to reason, argue, and persuade are worth learning."

Both the Middle States Association accreditation report and the Board of Visitors Special Committee recognized that midshipmen need better critical-thinking skills, which may be "the most important foundation for the leaders of the future." Too few midshipmen ever get to understand that a higher academic bar does exist. They might never choose to attempt it themselves, but they should know what it looks like. Or, as Admiral William Crowe, former Chairman of the Joint Chiefs of Staff, wrote in his autobiography, "excellence, intellectual rigor and curiosity are things that all officers should appreciate." By implication, so should all midshipmen.

Lesson Five: Allow Them to Experience Failure

With almost universal admiration, the Academy faculty and administration I interviewed identified a midshipman skill: Ability to focus on what needs to be done to get by in the short run. Some excel at it, achieve high grades, and thrive. Others learn to survive. What virtually none of them get to do—and remain at the Academy—is experience failure, except in the most trivial ways. Plebe year is about the transition from civilian to military life, pressure, and learning to respond to it. It is, as the Commandant put it, "Learning to keep many balls in the air successfully."

In many ways, however, the challenges of Academy life decrease as one progresses. Once a midshipman survives plebe year, the opportunity to fail—and graduate from the Academy—ceases. Learning from failure and preventing its recurrence is a life skill that all midshipmen need to learn and master. The headlines are full of criticism about the military's zero-defects mentality. Whether wittingly or not, the Academy is sending its young officers into the fleet with precisely this mind-set and experience.

The Ivy League has done rather well in promoting responsible risk-taking and tolerating failure by deemphasizing grades—not deemphasizing excellence or competition. Rather, they encourage students to try courses outside their area of strength without fear that it will hurt their grade-point average and chance of admission to top professional or graduate schools. In contrast, the Academy has an almost ante-deluvian reliance on grades, class standing, and color competition.

If, as a number of officers and midshipmen confided, very few people really care about these metrics, why spend so much time and energy on them? Tradition? Inertia? Public relations implications? Since virtually ever officer I interviewed referred to the zero-sum game of limited time at the Academy as the principal reason why desirable policies and experiences are not added, the lesson seems clear: Get rid of the unimportant stuff, and use that time more wisely.

Lesson Six: Recognize That Cynicism Is a Problem

Recently, I read John Feinstein's marvelous book, *A Civil War* (New York: Little, Brown, 1996), about a year in the life of the Army-Navy game. I came away from it with a sense that Feinstein had enormous respect and admiration for the midshipmen and cadets; but that he also found the academies to be rather cheerless places. When I interviewed Feinstein, he confirmed my interpretation: "They're great places to be from, but pretty unhappy places to be at," he said.

Virtually every Academy officer I asked said, no, the Academy was not an unhappy place; but they admitted that the midshipmen are tremendously cynical. Cynical about what? Without exception, every midshipman and most of the graduates I spoke with—including the most successful and the most gung-ho—expressed cynicism about many of the regulations and restrictions under which the midshipmen live. While 95% of midshipmen (according to Academy surveys) accept the premise that the Academy should be a "high-pressure, high-stress environment," they learn quickly to distinguish between what is mission-critical and what is not. Whether it is required attendance at "spontaneous" pep rallies or student guards at football games to prevent midshipmen from visiting the stands on the opposite side of the field, they experience and put up with policies that make little sense to them.

That midshipmen are cynical is not surprising; it is just unfortunate. Most can accept the deprivation of liberties their civilian friends take for granted. What they find more difficult to accept is that four years of in-extremis rules are necessary to make them better officers.

More than a few career officers expressed the sentiment that a tough plebe year combined with a more liberalized upper class experience would have been more valuable. Indeed, when I have suggested higher standards coupled with less "Mickey Mouse," heads begin nodding and suggestions begin flowing.

Perhaps inadvertently, the Commandant lent credence to this thesis. Admiral Roughead shared with me an example of midshipman input into a problem affecting first classmen: how to deal with a shortage of parking spaces. He was extremely pleased with the outcome and cited midshipmen's satisfaction with the result. That should have been a clarion call. What would happen if midshipmen were asked to scrap the reg book and help devise a plan to achieve the Academy's mission and improve morale?

Lesson Seven: Broadcast the Academy's Message

The Academy is tough, and it is not for everyone. But with the cost of an Ivy League education exceeding $100,000, the Academy should have broader appeal to more of the kids who wind up at top private colleges. The Ivies are surprisingly aggressive and fairly sophisticated in their marketing. With the cost of attending on the rise, the selling job they have to do—particularly to parents—is a significant one. The Academy needs to broaden its communication efforts, reaching out to the pockets of the United States—many of them in affluent areas—and "sell" the Academy and the military.

Lesson Eight: Broaden the Admission Pool

One reason the Academy may be misfiring on one cylinder is its use of the Strong Campbell Interest Inventory as part of the selection process. Career counselors long have used this instrument but recognize that it has real limitations. The most significant is that it is weakest when scored and interpreted en masse by a computer. It is most useful when used as an individual consulting tool, precisely the opposite of the way the Academy uses it.

Whether or not the Strong Campbell is an adequate or much-too-relied-upon tool for predicting a candidate's likelihood to stick with a naval career can be debated. What cannot be ignored is the fact that the current selection process is producing a wrong fit 25% of the time.

The lesson the Academy should learn from the Ivy League is not to seek out the "well-rounded" kid. Rather, like the Ivy League, it should broaden its admission pool to create the "well-rounded class"—some super scholars, some athletes, some gung-ho types, and some who will be the "glue of the class," as Harvard puts it.

Will this lead to a higher graduation rate and comparable service retention? Only time will tell. The Academy has the leverage to select from a large, highly qualified pool of applicants. The *Time Magazine/ Princeton Review* guide gives the Academy a "selectivity ranking" of 99, including it among the most selective colleges in the country. In the Ivy League, only Harvard, Yale, and Princeton receive a 99; Cornell gets a 97, and the remaining Ivies score 98. With one out of four midshipmen leaving before they graduate, the Academy needs to experiment.

Lesson Nine: Four Years Might Not Be Enough

The Academy's faculty and administration are both justifiably proud and relatively candid: They would like to improve both the academic experience and the professional training. The question—and it is legitimate—is how to do it in four years. The answer may lie in increasing the Academy program to five years. (One of the most successful Ivy League programs is the combined six-year bachelor's medical school degree offered at Brown.) While several faculty and administrators expressed a hypothetical desire to increase the length of the Naval Academy experience, it always was accompanied by a sigh and expression of political impossibility. Surprisingly, however, that political impediment is not one that I have heard from any politicians.

The Bottom Line

Can more rigor, broader experiences, and improved morale be wedged into a program that often seems bursting at the seams? Can the energy and creativity that now goes into "beating the system" be channeled into more productive pursuits? Or is the Academy experience truly a zero-sum game? In the current environment—with its reliance on structure and imposed discipline—it probably is. But change any one parameter, adopt any one of the Ivy approaches, and the opportunities for improvement are substantial.

The underlying question is whether it is worth the effort to find out. Perhaps this 1996 reminder from the late Admiral Jeremy Boorda is the most compelling reason why the Academy should consider these reforms seriously: "An organization that does not realize it can improve, is an organization designed to fail."

Changing the way the Academy treats midshipmen—with much of such treatment ingrained in tradition—demands tremendous faith in these young people. I have that faith. I hope Navy leaders have it too.

EDITOR'S NOTE: *For a summary of a companion piece outlining what the Ivy League schools can learn from the Naval Academy, see the U.S. Naval Institute website at www.usni.org. For the full text, see* Brown Alumni Magazine *at: www.brown.edu/Administration/ Brown_Alumni_ Magazine/index.html.*

Mr. Cohen is a Managing Director at Scholastic, the children's book publisher. He is the author of four best-selling books, all of which focus on education, and has lectured extensively on campuses from Fordham to Stanford.

"Why We Teach Leadership and Ethics at the Naval Academy"

13

Captain Mark N. Clemente, USN

U.S. Naval Institute *Proceedings*
(February 2000): 86–88

NOT SURPRISINGLY IN THIS ERA of intense scrutiny of public institutions, the ethics and leadership curriculum at the U.S. Naval Academy has attracted media interest. The majority of the recent coverage has been accurate, but a few editorials have missed the mark and misrepresented the program.

It also is interesting to note that while newspapers today highlight recruiting problems in all the services, the Naval Academy remains oversubscribed. The best of Generation Y, it appears, still want to come to the Naval Academy, and they're bringing mean Scholastic Aptitude Test scores above 1300.

When I came back to teach leadership and ethics, what I found surprised me: the Naval Academy is much better than when I entered almost 25 years ago. The midshipmen are in better physical condition, their collective grade point average is higher, and character development is taken much more seriously.

I believe that part of the Naval Academy's attraction to top-quality young men and women is its unambiguous mission:

To develop midshipmen morally, mentally, and physically and to imbue them with the highest ideals of duty, honor, and loyalty in order to provide graduates who are dedicated to a career of naval service and have potential for future development in mind and character to assume the highest responsibilities of command, citizenship, and government.

In fact, the Naval Academy does produce leaders of character prepared to serve in the combat arms of the Navy and Marine Corps, and what follows is a factual account of the program as exists today. Leadership development is woven into virtually every aspect of a midshipman's four years here. Leadership emphasis permeates the institution—the classroom, in Bancroft Hall (their quarters), on the parade ground, during summer training cruises, and on the athletic "fields of friendly strife." It exists in places like Memorial Hall, John Paul Jones's Crypt, and the chapel—places that recall the sacrifices of those Americans who have gone before.

The journey to become a leader worthy of leading the sons and daughters of America begins the minute a new fourth classman walks through the gates to undergo "plebe summer," six weeks of challenging, stressful military indoctrination. While some have prior enlisted experience, the majority are scant weeks removed from their high school graduations. The training includes a rigorous physical conditioning program designed to get the new midshipmen in top physical condition, create an understanding that physical fitness is a necessary attribute of a warrior, and instill in them the desire to pursue a lifetime of physical fitness. We believe in the adage that "You have to learn how to follow, before you can learn how to lead," and the new midshipmen are exposed daily to hand-picked members of the second and first classes who serve as their leaders. Plebes get their first introduction to the Honor Concept, and hear the Chief of Naval Operations and the Commandant of the Marine Corps reinforce the core values of the naval service.

Leadership development is expanded and refined over the course of a midshipman's education and training. With each succeeding academic year, midshipmen are called upon to assume greater and more significant responsibilities. Of course, leadership development does not end with graduation. Rather, just as we prepare midshipmen for a lifetime of learning and fitness, we also prepare them for a lifetime of leadership development and achievement.

The Naval Academy leadership development program requires observation, education, reflection, and practice. It gives midshipmen a more sophisticated understanding of the fundamental principles and core values that should guide a leader, and provides an intellectual grounding in those principles and values without forgetting that we are preparing these young men and women to lead in combat.

Although leadership is embedded in nearly every activity at the Naval Academy from the positions held by the "stripers"—those who hold official leadership positions within the brigade—to extracurricular activities, my focus here is primarily on the formal leadership curriculum in the Department of Leadership, Ethics and Law.

The classroom experience blends leadership theory and practice to provide each midshipman a solid intellectual framework and an understanding of key leadership principles. Experience is a great teacher, but leaders must be able to think through situations they have never experienced. Midshipmen also learn the basics of human and group behavior. The study of these disciplines is fundamental, but it also is important to note that the Naval Academy presents this information in the context of leadership in the profession of arms.

To keep the focus on military leadership, many of the classroom case studies and examples that bring theory to life come from combat and the fleet. The military instructors who teach leadership have served successfully in the fleet or Fleet Marine Force and have current and varied operational experience—naval aviators with combat experience in Desert Storm, Navy and Marine Corps officers recently back from the

Balkans, the Persian Gulf, or the western Pacific. Commanding officers or others recently in command frequently visit to talk with midshipmen in individual classrooms. These warrior role models bring insights from the real world to midshipmen consumed by the daily grind, helping them to remember why they came here in the first place.

After the initial plebe summer indoctrination, the new fourth classmen take a course titled Leadership and Human Behavior. A newly revised course taught by officers, it explores the effects of individual human behavior on leadership and being led. It enables our aspiring leaders to examine their own motivation and leadership skills, and covers a wide array of topics from personality to perspective; from how we learn to how we react and make decisions under stress. The seminar style of classroom discussion is peppered with case studies from the fleet to make the theory resonate with the students

In their second-class, or junior, year, the midshipmen take an advanced leadership course that examines the leadership process through the dynamic interaction between the "leader, the follower, and the situation." Elements of group dynamics and the skills required to motivate organized groups to accomplish a mission are the focus. Case studies include Battle of the Chosin Reservoir during the Korean War, Oliver Hazard Perry at the Battle of Lake Erie, the Iraqi missile attack on the USS *Stark* (FFG-31), and the mine strike of the USS *Samuel B. Roberts* (FFG-58).

These two academic courses complement the practical leadership lessons our young men and women learn in Bancroft Hall, on summer cruise, on yard patrol and sailing craft, and in athletic competition. In addition to interaction with leadership instructors in the classroom, an important part of our leadership development program are the day-to-day engagements midshipmen have with the top-quality Navy and Marine Corps officers and senior enlisted men and women assigned to the Naval Academy faculty and staff. Senior enlisted in each company have significantly enhanced the Bancroft Hall experience.

Why should a future combat leader study ethics? Doesn't morality come from our family and religious experience? We often hear these questions. Our profession is one of arms, and it is this awesome power—the power to take life—that makes it imperative that our graduates understand the gravity, responsibility, and nobility of this profession.

Vice Admiral James B. Stockdale, U.S. Navy (Retired), who was awarded the Medal of Honor for his leadership as a prisoner of war during the Vietnam War, encouraged the teaching of ethics and philosophy to military professionals. The University of San Diego hosted a James Bond Stockdale Leadership and Ethics Symposium last winter, and hopes to establish a chair in his name. Here's how the university described him at the symposium:

> Admiral Stockdale has borne witness with his entire life to the importance of studying, understanding, and applying ethical principles to one's profession and the conduct of one's life. His efforts in his early military career and in his first graduate studies to learn the basic premises of philosophical ethics served him well during the terrible hardships and deprivations he suffered while a prisoner of war from 1965–1973. His knowledge of the philosophy of Stoicism, especially as taught by the Roman philosopher, Epictetus, provided him not only a framework within which to bear with his own ordeal, but also a model for his role as senior officer and leader of all the prisoners confined in the notorious Hoa Lo prison. Admiral Stockdale led by example, as well as by precept. . . .

A course in professional military ethics at the Naval Academy concludes a semester-long exploration of military case studies and military applications of basic philosophical principles with consideration of Vice Admiral Stockdale's experience as a prisoner of war. The midshipmen read and discuss Epictetus, the philosopher he credited with providing him inner strength during his ordeal, alongside his own account of his military experiences. In November 1999, he spoke to the entire third class about his experiences over nearly four decades of service.

During their third class (sophomore) year, the midshipmen explore these ideas in a required course entitled Ethics and Moral Reasoning for the Naval Leader (NE 203), which emphasizes that one is responsible for the consequences of one's actions as well as one's intentions. We are not teaching situational ethics. Ethics and Leadership here is not a touchy-feely "I'm okay, you're okay" curriculum. Absolutes exist, and what is right, the truth, many times is known. For those situations, we expect compliance and demand accountability. As Dr. Nancy Sherman, the Naval Academy's first Distinguished Chair of Ethics, pointed out in these pages in "Ethics for Those Who Go Down to the Sea in Ships" (see *Proceedings*, April 1999, pages 87-89), the motivation behind why someone follows the rules, and the accompanying emotions, is what we are trying to tap into. If the only reason someone avoids crime is fear of punishment, we need to lift their sense of moral responsibility; if the only reason they do something is to please somebody—that is not good enough. We want leaders who have internalized the values we cherish and promote. We fail if they view leadership and professional military ethics as merely legalistic or contractual.

But it is the need to prepare midshipmen to deal with those cases that are not so clear-cut also that is the basis for today's curriculum. We want them to graduate with a true understanding of timeless principles, so whenever they encounter situations where the rules have yet to be written, or where conflicts of duty are encountered, they can dig into their conscience and apply the critical thinking skills they learned during their four years at the Naval Academy and do the right thing—and for the right reasons. We want, in the words of John Paul Jones, someone with " . . . a liberal education, and the nicest sense of personal honor."

The Naval Academy's ethics course is unique precisely because it is not "just another academic course" taught by civilian faculty. It is taught by military officers. The requirement for serving as an ethics instructor in our course is, at minimum, having attained the rank of commander/lieutenant colonel. Civilian faculty, also, are involved in, and committed to, our leadership and ethics program. We are proud of the 150-year tradition at the Naval Academy of having a substantial cadre of distinguished and dedicated civilian faculty who teach alongside their military colleagues. Civilian faculty specializing in philosophical ethics provide insight into the history and meaning of the fundamental values of Western culture. They provide the theory in our ethics course that the military instructors translate into practice. The Naval Academy faculty has also been augmented by two endowed chairs: the visiting Distinguished Leadership Chair, Admiral Hank Chiles, U.S. Navy (Retired), and our visiting Distinguished Chair in Ethics, Dr. Douglas MacLean, from the University of Maryland Baltimore County.

Military officers provide the grounding in practical experience that puts those ideals to the test. Civilian faculty, especially those new to the Naval Academy and its rich traditions, describe this daily interaction with senior military officers as among the most challenging and rewarding experiences of their academic careers. We want our graduates to reason morally in peace and in war.

Our ethics course, therefore, is a critical survey of the major moral theories in the Western tradition, each of which has had something to say about what makes actions right and about the source of the authority of morality. Midshipmen study classical utilitarianism and duty ethics along with divine command theory, natural law theory, and virtue ethics. They read and discuss the original work of philosophers from Aristotle through St. Thomas Aquinas to John Stuart Mill and John Rawls.

The classical Greeks were interested in what sort of life is the best life for human beings. Later, St. Thomas Aquinas incorporated many elements of Aristotle's view into Christian natural law theory, which holds that the authority of moral precepts comes from God and is revealed in natural human dispositions. For a variety of reasons, 17th century philosophers took a different approach. They sought to articulate a moral theory expressed in terms of rules or principles and one that would resonate with people of a variety of faiths and with people who had no religious commitments. Bentham and, later, John Stuart Mill developed classical utilitarianism. According to this view, human welfare is of prime moral importance, and the right thing to do is maximize this welfare; they urged us to focus on the consequences of our actions.

It is clear, however, that we care about more than the consequences of our actions. The intentions on which we act are also of moral significance, and Immanuel Kant developed his duty ethics with this in mind. As with any serious academic discipline, there is room for disagreement about which of these approaches to morality is complete and correct. But midshipmen can learn from each of them.

The ethics syllabus reflects a military case-study oriented approach to the presentation of the course material that parallels the standard approaches to teaching professional ethics in other professions such as medicine, law, and business. The extensive use of a variety of cases drawn from actual situations is intended to assist midshipmen in understanding the practical applications to military life of the moral principles and ethical theories examined during the course.

There are those who criticize the notion of learning leadership or ethics through the study of theory. We reject this. Practice without theory and reflection is, at best, risky, and at worst a shot in the dark. On the other hand, theory without practice is hollow. While the rate of change in mankind's knowledge of himself and the universe has increased, there remain timeless principles surrounding human behavior that we must try to understand. If Sun Tzu still makes sense after 2,500 years, we ignore theory and reflection at our peril.

We're proud of our program, and we believe that the quality of our graduates—our only measure of success—speaks for itself.

Captain Clemente, an F-14 radar intercept officer, is the Chairman of the Leadership, Ethics and Law Department at the Academy, and teaches ethics to midshipmen. He commanded Fighter Squadron (VF)-213.

14 "U.S. Naval Academy Is Balanced"

Vice Admiral James F. Calvert, USN (Ret.)

U.S. Naval Institute *Proceedings*
(October 2003): 26

AN ARTICLE IN THE 14 JULY issue of *The National Review*, "Babylon Comes to Sparta" by John J. Miller, has raised considerable attention on the web. Because I am sometimes accused of having started all this trouble with the majors program introduced during my tour as Superintendent, I believe some comment is in order.

First, there was no doubt that the old lockstep curriculum where all midshipmen took an Electrical Engineering major had to be changed. The brigade was becoming more inclusive, to use today's word for it, and the competition for top-notch candidates was becoming ever more severe.

A great deal of work, including assistance from three college presidents on my Academic Advisory Board, went into the creation of the majors program. Admiral Thomas Moorer, then Chief of Naval Operations, was consulted many times in the process, and we had his complete approval.

Within a few years both the U.S. Military Academy and the U.S. Air Force Academy adopted similar programs.

The Navy is the most technical of the services, not only because of nuclear power, but also because of its heavy dependence on electronics and computers in its ships and planes. Nevertheless, it is necessary to have at least some officers who have concentrated on less technical matters. Thus there are majors in History, English, and Political Science while the other 16 are in engineering, math, or science fields.

The main thrust of the criticism, however, is about the ethics and leadership programs that somehow are seen as destroying the warrior spirit.

We have to recognize that these programs were initiated in the wake of the electrical-engineering scandals of the early 1990s. Admiral Charles Larson, when brought in for his second tour as Superintendent, was expected to, and did, institute programs designed to change the basic ethical standards of young people in their late teens and early 20s—not an easy task.

Plebe summer surveys, given with a strict requirement for anonymity, revealed a disturbingly lax attitude about cheating and lying—an attitude that stemmed from the high school culture they had just left behind. In these anonymous surveys more than three-quarters of the incoming plebes stated that they had cheated "frequently" in high school. Anonymous surveys at West Point showed similar results.

I was so disturbed by these facts that I spent some time quizzing contemporary high school students whose opinions I believed I could trust. Almost unanimously they agreed that it was true and that "the worst offenders were students who were getting all As." Perhaps they wanted those high grades to go to a competitive university such as Harvard or Stanford or Wellesley—or Annapolis or West Point.

In any case, it was clear that an effort had to be made to alter these views at a time of life when such changes do not come easily. A significant effort was made at the Naval Academy with the establishment of whole academic departments with these objectives.

There are, of course, flaws and shakedown problems in these programs, but the fact remains that the service academies (and, I believe, particularly West Point and Annapolis) are leading the way among four-year accredited universities in efforts to provide vigorous programs of these kinds in their curricula.

I have attended, along with several classmates and other senior Naval Academy alumni, several of these classes in ethics and leadership and have examined their text material. I believe that they are aimed in the right direction and that they will improve as the years go by. In addition, company officers are well aware of the new emphasis on character and the honor system and these officers are an important adjunct to the effort.

As for the warrior spirit, I do not believe we need look farther than the recent Iraq War for proof that our young graduates, both Navy and Marine Corps, have plenty of it. The performance of young Marine officers, the Navy Seals, and our naval aviators should make us all proud of our Academy.

Lastly, our civilian faculty, which has long been a proud tradition at Annapolis, is a vital part of the fine education we offer. If any institution of higher learning thinks that it can attract first-rate faculty members today with all of them possessing hard-boiled right-wing opinions, it is not facing reality. Today's faculty members were mostly in college during the 1960s and 1970s where they got a full dose of liberalism that most of them will only slowly, if ever, lose.

The balance that must be provided at a service academy is to have a carefully selected officer faculty (about half of the total faculty at Annapolis) and even more carefully selected company officers. It is the task of these officers, which I believe they take seriously, to make it clear that the purpose of the Navy and the Marine Corps is the defense of our nation and this often cannot be done with other than brute force and violence.

I believe we are most fortunate to have our midshipmen exposed to this balance between academia and military reality—and we should not get wobbly on it because of criticism by either radical right or left wingers.

Vice Admiral Calvert was Superintendent of the U.S. Naval Academy, 1968–1972.

15 "A Choice for the Oath—Game of Thrones and the U.S. Naval Academy"

Lieutenant Kelly Maher, USN, and
Lieutenant Michael Maxwell, USN

U.S. Naval Institute *Blog* (29 May 2015)

THE FIFTH SEASON of the HBO hit-series *Game of Thrones* is here! I'm excited, as are millions of die-hard fans across the country. To prepare for the imminent launch, I re-watched all four of the previous seasons, episode by episode. In that first season, an interesting event takes place, where a young man, Jon Snow, is given his duty assignment. He is about to take an oath to serve for life in the Night's Watch. He has prepared for years to be a Ranger—a fighter and swordsman. Instead he is assigned as a Steward. Jon Snow is crushed. He hasn't taken the oath of service yet, and he contemplates leaving the Night's Watch to avoid a life of inglorious servitude as a steward. His friend Sam convinces him to stay, reminding him that service is about more than his own selfish desires. Jon Snow takes the oath later in the episode.

It brought me back to my own service selection. I dreamed for years and years of becoming a Marine Corps Officer. At the Naval Academy that fateful day in November of 2009, I received troubling news—I had been selected to become a Surface Warfare Officer. Over the years since I have often been asked if I wanted to become a SWO. My standard reply is that it was one of my top six choices. The humor gets me through the moment, and the conversation moves on.

149

I'm working now at the Academy, preparing to take over as a company officer this summer, just in time for the Plebe Class of 2019 to arrive for I-Day. I am a proud Surface Warfare Officer and I wouldn't trade it for the world. I have been to more overseas ports than I can count over two deployments, have navigated tens of thousands of miles at sea, and served with some of the bravest, smartest and most loyal Sailors the world has ever seen.

Much of the conversation within the walls of the Academy frequently turns to an age-old symptom of the institution—cynicism within the Brigade. Midshipmen sometimes complain that they aren't treated like future naval officers and that they aren't doing real work to prepare themselves to become the leaders of those fine Sailors and Marines. "I'm going to fly jets, why do I need to learn about buoy systems in the Western Rivers" is just one example. In teaching leadership on the yard, we strive for every class to fight that mentality, to prove to these young Midshipmen that their training is exceptional and that they will be well prepared to lead upon commissioning. Sometimes I fear that we aren't doing enough, that the Midshipmen are right, and that we are sending our future junior officers to the fleet without the preparation needed to fulfill their duties. For the graduating Midshipmen, winter is coming, and many aren't ready to handle a sword.

I don't know entirely where the cynicism comes from, but I have a theory. Everything for these Midshipmen centers around one key event—service selection. Competition is fierce within the Brigade. Classmates vie for position and jostle for rank as if they were in Westeros, the fictional land of Game of Thrones. There are only so many slots for SEALs, Marines, Submariners, Aviators, and today even SWOs. Midshipmen study diligently to get good grades, so that their order of merit is high enough to get the service selection they want. Many spend more effort on good grades to earn that service selection, but in doing so disregard the very skill sets required to be successful naval officers—pro-knowledge is an afterthought and weighted minimally when compared to calculus and chemistry. The drive for service assignment goes beyond academics, of course. They perform with vigor on the PT fields to notch themselves up for the same purpose. Those wanting to become Marines join the Semper Fi Society, those seeking to become SEALs test themselves and compete against their classmates in arduous screeners.

That day in November, the Firsties learn their fates. Most are over-joyed—a good thing, no doubt. A few feel despair. These are the ones we should worry about. These are the examples that feed the cynicism —working hard may not be enough. These are the few who enter the fleet sullen, downcast and doubtful. These are the ones most unprepared for their future roles, having spent all of their efforts learning about fire team movements and squad assaults instead of honing their shiphan-dling skills on the YPs. These are the few who, in my opinion, are the least likely to commit themselves to a full career of service and will leave at the earliest opportunity.

Even those who earn their top choice are too hastily prepared for the training to be effective, meaning that the Chief's Mess, Department Heads, and Commanding Officers are burdened with teaching junior officers skillsets and professional knowledge they should have mastered at the Naval Academy. The unit leadership should be focused on advanced training—on defeating multiple threats simultaneously, mastering com-plex engineering systems and conditioning our new Ensigns and Second Lieutenants to become outstanding naval leaders. Instead, they are too busy teaching standard commands, basic maintenance protocols and general military socialization.

What if we changed something? What if we moved service selection to the end of Youngster (sophomore) year? By that time, Midshipmen will have been able to establish their grades, competed in screeners, etc., at least enough for the Academy to choose wisely between them. We could move PROTRAMID, a fleet-wide round-robin experience to expose the

Midshipmen to the various communities to the end of Plebe year, just like the NROTC currently does, to allow our new Youngsters the opportunity to see what fits them best. Most Plebes know what they want to service select before they climb Herndon, while the rest of the class would have another year to weigh the decision.

This change has several notable benefits. First, it eliminates competition amongst classmates during their junior and senior years, allowing for greater opportunity to hone leadership and professional skills in Bancroft. Second, it provides two full years, instead of a meager four months, for Midshipmen to hone their practical skills, affording them the chance to excel in tactical and technical competence from day one in the fleet. Marine selectees will have two years to practice ground tactics. Aviators have two years to pass IFS, easing the burden on Pensacola and the subsequent stashing of officers on the Yard until flight school begins. SWOs can master navigation and shiphandling before setting foot on the bridge of a destroyer. Third, if we rearrange the course loads, we can eliminate the cynicism that arises from taking courses that Midshipmen see as irrelevant, such as Marine wannabes having to struggle through seamanship and navigation courses. Fourth, and possibly most importantly, it allows Midshipmen a choice. They now know what they will be doing for their careers and if those few who don't earn what they want choose to leave before signing their commitment papers the next Fall, the fleet will benefit from a drop in uncommitted and unenthusiastic naval officers. If a Midshipman is so disappointed in his or her service assignment, he or she doesn't have to come back to poison the well back in Bancroft, or worse yet, carry that attitude into the fleet. Furthermore, by encouraging choice, we disrupt cynicism about being treated like children—a Midshipmen knows full-well what he or she is getting into when they sign on the line which is dotted.

Secretary of the Navy Ray Mabus recently spoke to the brigade about a number of institutional changes aimed at improving talent management and retention. He mentioned that the Academy is already moving towards a system that seeks to match talent to title and is less dependent on class rank. He and his staff clearly understand that change is needed, not only for its effect on the yard but also downrange in the fleet. This proposal provides an avenue for that change, even if it is one of many. In combat, a coordinated simultaneous time-on-top attack is always preferred to a slew of single efforts and I believe that changing the timeframe for service selection is a key weapon in the fight against complacency and cynicism to ensure we maintain the highest level of combat readiness throughout the fleet. Even if our ships rust and our airframes crack, our people must remain sharp and steadfast.

Choice is nobody's enemy. While I don't have the same flowing locks and sword skills as Jon Snow, I empathize with his decision. I didn't want to be a SWO, at least not initially, but my call to service outweighed my selfishness. I figured that if I was going to be a SWO, I would try my damndest to excel at it. Under this proposed change, there will still be plenty of disappointed Midshipmen who put their country before themselves and will accept what they earned with grace and humility. They will remember that service and leadership are what count, not the uniform they wear or the devices on their chest.

16 "Shake Things Up on the Yard"

Senior Chief Jim Murphy, USN (Ret.)

U.S. Naval Institute *Proceedings*
(July 2015): 14

SECRETARY OF THE NAVY RAY MABUS is leading change. He has made or recommended modifications in everything from general military training to paternity leave. He is influencing changes to career-intermission programs, training strategies, career opportunities for women, promotion policy, and physical fitness.

Secretary Mabus is shaking things up across the fleet, and nothing appears off the table. Perhaps he will consider disturbing the peace at the U.S. Naval Academy.

A recent USNI Blog post by two midshipmen indicated a desire among the brigade to be challenged intellectually and for academics to be more closely tied to skills needed for success in the fleet ("Challenge Us! Two Midshipmen's Plea for Comprehensive Problem-Solving in the Classroom," by MIDN 2/C Scotty Davids and MIDN 2/C Charlotte Asdal, 2 June 2015). Those are mature and reasonable demands but just two of several changes that might benefit midshipmen.

Naval Academy leadership has changed little over the past several generations, and significant progress will not occur by perpetuating old strategies. One area the secretary should consider changing is selection criteria for the Superintendent. This is not a criticism of the current Supe or his predecessors, but changing the *type* of leader selected to head the Academy would be appropriate.

For the past 134 years, the Naval Academy Superintendent has been a male graduate of the institution. Neither of those traits is a qualification, but both have been shared by the past 51 Supes since November 1881. Assigning an otherwise qualified leader with a different background would lead to change throughout the Academy.

Graduating from the Naval Academy should not be a discriminator for selection as Superintendent. The Navy benefits from a number of qualified flag officers who graduated from civilian educational institutions and a growing population of female admirals. Additionally, the Marine Corps has officers with similar backgrounds, and a Marine general could, and occasionally should, lead the brigade, faculty, and staff.

The services also have a number of flag and general officers with experience in education and training. For example, Rear Admiral Martha Herb is not an Academy graduate but is a warfare-qualified doctor of education. A reservist, she has relevant experience as the current Director of the Inter-American Defense College. She is but one example of a leader that would signify revolutionary change in Academy leadership; no specific recommendation should be inferred.

There are likely similar candidates. If not, that itself is a problem. The services routinely educate officers in specialties such as engineering, strategy, finance, and logistics. Education and training are no less important to fulfilling statutory responsibilities, and some number of officers should be afforded opportunities to develop appropriate knowledge and experience.

A proven leader qualified in a warfare discipline and with a background in the science of learning would seemingly make a great Superintendent. Although the academic dean and others in his department are experienced educators, there would be no harm, and potentially great benefit, in the leader of an educational institution having similar qualifications. And it does not matter where he or she earned their undergraduate degree, or whether they have served as a sailor or Marine.

An honest look at recent history and the requirements of the modern services will show that a diversity of educational background and leadership experience among those who lead the Naval Academy is needed. Change like that requires a catalyst. Secretary Mabus is a catalyst elsewhere, and if he is willing to break tradition in Annapolis, he could deliver academic and leadership-development changes that midshipmen and the naval services desire and deserve.

The preceding statements are heretical to many. In the June *Proceedings*, retired Coast Guard Captain R. B. Watts wrote, "The time for heresy is now." While he was "Advocating Naval Heresy" in our thinking about threats, strategies, and platforms, a broader application is appropriate and may be necessary to impact the areas he highlighted.

Routinely assigning the same type of leaders to the Naval Academy is not a recipe for driving significant and positive change. A diversity of educational background, operational experience, and chosen service would be. Assigning a non-graduate, a woman, or a Marine would be revolutionary. When the time comes, selection of the next Superintendent will present an opportunity to shake things up on the Yard.

Senior Chief Murphy retired from the Navy after 21 years of service. His June 2012 *Proceedings* column, "Bring Back Humility," was included in the recently published *Naval Leadership*, a volume in the U.S. Naval Institute Wheel Book series.

Epilogue: "The Heartbeat of a a Great Nation"

Jack Sweetman and revised by Thomas J. Cutler

(Selection from *The U.S. Naval Academy:
An Illustrated History*, 2nd Edition,
Naval Institute Press, 1995): 262–63

. . . BY MAY OF 1990, academy morale was seriously depressed. The commissioning week festivities were marred that year by the dominance of the academic and gender issues in the news media. *The Annapolis Capital* that week was filled with stories, commentaries, and letters to the editor about the problems in the Yard. Some called for Hill's [Superintendent] removal, others suggested that the matters had been blown out of proportion. Academic freedom, military discipline, women's rights, and the importance of tradition had become battle cries in the increasingly discordant dialogue. One letter to the editor in the Sunday edition sought the high ground between the warring factions:

I have just returned from watching what will be my last parade as a member of the staff of the U.S. Naval Academy (I am retiring from the Navy July 1).

As I watched those bright young faces march past me, I was deeply moved as I always am when witnessing this unique spectacle. But this time there was a bittersweet taste in my mouth, partly because I am leaving and partly because the Naval Academy is under such scrutiny these days.

I welcome that scrutiny because in order for an institution to remain great it must never become complacent. Greatness can be momentarily achieved by a single act, but it can be sustained only by a continuous willingness to self-examine and re-evaluate. There is no dishonor in having weaknesses, only in ignoring them.

But as we witness these times of introspection and public scrutiny, I would ask that we never lose sight of what we are about. As I feel the thunder of drums on Worden field, I sense that I am feeling the heartbeat of a great nation. And as I hear the echoes of bugles reverberating against the monuments that proliferate this campus we call "The Yard," I am reminded of the greatness that has gone before us and will continue to emerge from within these walls despite the problems we are experiencing and working hard to solve.

My tour of duty is coming to a close, but as I look out through misted eyes across that sea of blue and gold uniforms and see the sun gleaming off highly polished eagles and anchors, I know that the Navy will go on without me and that the Naval Academy will weather this storm and continue to be the heart and soul of the Navy that it has been for nearly a century and a half.

In these important weeks to come, let us all take pride in the fact that we are a nation where such scrutiny can take place; let us open our minds, correct our shortcomings, and learn from our mistakes; but let us never lose sight of the greatness which we are trying to make better.

INDEX

SERIES EDITOR

THOMAS J. CUTLER has been serving the U.S. Navy in various capacities for more than fifty years. The author of many articles and books, including several editions of *The Bluejacket's Manual* and *A Sailor's History of the U.S. Navy*, he is currently the director of professional publishing at the Naval Institute Press and Fleet Professor of Strategy and Policy with the Naval War College. He has received the William P. Clements Award for Excellence in Education as military teacher of the year at the U.S. Naval Academy, the Alfred Thayer Mahan Award for Naval Literature, the U.S. Maritime Literature Award, the Naval Institute Press Author of the Year Award, and the Commodore Dudley Knox Lifetime Achievement Award in Naval History.

The **Naval Institute Press** is the book-publishing arm of the U.S. Naval Institute, a private, nonprofit, membership society for sea service professionals and others who share an interest in naval and maritime affairs. Established in 1873 at the U.S. Naval Academy in Annapolis, Maryland, where its offices remain today, the Naval Institute has members worldwide.

Members of the Naval Institute support the education programs of the society and receive the influential monthly magazine *Proceedings* or the colorful bimonthly magazine *Naval History* and discounts on fine nautical prints and on ship and aircraft photos. They also have access to the transcripts of the Institute's Oral History Program and get discounted admission to any of the Institute-sponsored seminars offered around the country.

The Naval Institute's book-publishing program, begun in 1898 with basic guides to naval practices, has broadened its scope to include books of more general interest. Now the Naval Institute Press publishes about seventy titles each year, ranging from how-to books on boating and navigation to battle histories, biographies, ship and aircraft guides, and novels. Institute members receive significant discounts on the Press' more than eight hundred books in print.

Full-time students are eligible for special half-price membership rates. Life memberships are also available.

For a free catalog describing Naval Institute Press books currently available, and for further information about joining the U.S. Naval Institute, please write to:

Member Services
U.S. NAVAL INSTITUTE
291 Wood Road
Annapolis, MD 21402-5034
Telephone: (800) 233-8764
Fax: (410) 571-1703
Web address: www.usni.org